BY JOHN ELDER ROBISON

Look Me in the Eye

Be Different

Raising Cubby

Switched On

Switched On

Switched On

A MEMOIR OF BRAIN CHANGE AND EMOTIONAL AWAKENING

John Elder Robison

Foreword by Alvaro Pascual-Leone, MD, PhD

SPIEGEL & GRAU NEW YORK

2017 Spiegel & Grau Trade Paperback Edition

Published in the United States by Spiegel & Grau, an imprint
of Random House, a division of Penguin Random House LLC, New York.

Spiegel & Grau and Design is a registered
trademark of Penguin Random House LLC.

Originally published in hardcover in the United States by Spiegel & Grau,
an imprint of Random House, a division of Penguin Random House LLC, in 2016.

LIBRARY OF CONGRESS CATALOGING-IN-PUBLICATION DATA
Names: Robison, John Elder, author.
Title: Switched on: a memoir of brain change and emotional awakening /
John Elder Robison.
Description: New York: Spiegel & Grau, 2015. | Includes bibliographical references.
Identifiers: LCCN 2015014112 | ISBN 9780812986648 (print) |
ISBN 9780812996906 (ebook)
Subjects: LCSH: Robison, John Elder—Mental health. | Asperger's syndrome—
Patients—United States—Biography. | Asperger's syndrome—Patients—Treatment. |
BISAC: BIOGRAPHY & AUTOBIOGRAPHY / Personal Memoirs. | SCIENCE /
Life Sciences / Neuroscience. | MEDICAL / Neuroscience.
Classification: LCC RC553.A88 R6356 2015 | DDC 616.85'88320092—dc23
LC record available at http://lccn.loc.gov/2015014112

randomhousebooks.com
spiegelandgrau.com

Book design by Christopher M. Zucker

146119709

For Maripat,
who brought our family together with love.
And Mary,
who left us too soon.

I'm living at a peak of clarity and beauty I never knew existed. Every part of me is attuned to the work. I soak it up into my pores during the day, and at night—in the moments before I pass off into sleep—ideas explode into my head like fireworks. There is no greater joy than the burst of solution to a problem. Incredible that anything could happen to take away this bubbling energy, the zest that fills everything I do. It's as if all the knowledge I've soaked in during the past months has coalesced and lifted me to a peak of light and understanding. This is beauty, love, and truth all rolled into one. This is joy.

—DANIEL KEYES, *Flowers for Algernon*

Contents

Author's Note   xiii

Foreword by Alvaro Pascual-Leone, MD, PhD   xv

Prologue   xix

An Electrifying Proposal   3

The Value of Detachment, Circa 1978   13

Medical Magnets   26

Why Change?   31

Horsepower   35

Informed Consent   40

The History of Brain Stimulation   49

Mapping My Brain   57

The Night the Music Came Alive   67

Emotion   80

Singing for Ambulances   86

A Family Affair   96

Seeing into People   103

Hallucinations and Reality   115

Awakening   131

Science Fiction Becomes Real   144

The Zero-Sum Game   157

The Shimmer of Music   167

Aftermath   172

Nature's Engineers   184
Speech   191
A More Subtle Result   196
Different Kinds of Success   205
Rewriting History   211
Fear   216
A New Beginning   225
Tuning Out the Static   229
Mind Readers   238
A Death in the Family   246
Back in the Groove   255
Postscript: The Future   264

Afterword by Marcel Adam Just, PhD   277
Findings and Further Reading   283
Acknowledgments   291

Author's Note

SOME READERS of this story may ask if it's true. It is. This is a memoir that describes my participation in a series of brain stimulation experiments and what happened next.

Events and conversations were reconstructed as accurately as possible using notes, emails, and the collective memory of others who shared this remarkable adventure with me. While I put words in the mouths of various people, notably the physicians and scientists at Beth Israel Deaconess Medical Center, I do not have recordings or detailed notes of all these conversations and do not mean to imply that those were actually their exact words. They are my best recollection, relating my own perspective.

With that in mind, I asked the key people whose conversations and actions fill this book to read what I wrote and make sure I did not inadvertently misrepresent the things they said or did.

The doctors and scientists involved are all identified, as are most other people in the story. I've changed the names of a few other characters to disguise their identity, and made note when that was done, but in every case events are related as accurately as my memory allows.

I've done my best to avoid errors when explaining complex neuroscience, and I've asked the key scientists to review everything I've written for technical accuracy. I have incorporated their many corrections and explanations, and any remaining errors—and I'm sure there are some—are mine alone.

The journey described on these pages was made possible through the hard work and insight of Dr. Alvaro Pascual-Leone, MD, PhD, and the staff of the Berenson-Allen Center for Noninvasive Brain Stimulation at Boston's Beth Israel Deaconess Medical Center, a teaching hospital of Harvard Medical School. Without you, none of this would have happened.

JOHN ELDER ROBISON
January 2016

Foreword

Alvaro Pascual-Leone, MD, PhD

IN *SWITCHED ON*, John Robison has written a remarkable, engaging, and moving story that reminds me of why I first became a physician. In my practice, I try never to lose sight of this admonition attributed to Hippocrates, the Ancient Greek physician who was arguably the father of modern medicine: "If you have to choose between learning about the disease that a patient has, or about the patient who has a disease, choose always the latter." As a cognitive and behavioral neurologist, my mission is to help patients affected by various neurological and psychiatric conditions, including autism, epilepsy, stroke, Parkinson's disease, or drug-resistant depression. Modern medicine places immense importance on decision-making based on the most up-to-date science, and on accumulating specialized knowledge of disease, but the truth remains that clinical medical practice, at its core, should be about helping a person—a specific individual—get better. John Robison's story brings me back to this truth. It echoes through my head every morning when I walk through the doors of the Berenson-Allen Center for Noninvasive Brain Stimulation at Beth Israel Deaconess Medical Center, which I run and where much of the action in this book takes place.

The concept of "brain plasticity" refers to the ongoing capacity of the brain and the nervous system to change itself. Everything that we do, think, feel, and experience changes our brain. A stroke or a traumatic brain injury can affect brain plasticity, and plasticity may also be associated with such developmental disorders as autism. Increased brain plasticity may also potentially endow a person with unanticipated new abilities, as John appears to have experienced in this book. TMS, or transcranial magnetic stimulation, the intervention that John undergoes, provides a unique opportunity for us to learn about the mechanisms of plasticity, and to identify alterations in the brain's networks that may be responsible for a patient's problematic symptoms, and also for recovery.

Over the past three decades, TMS has become a valuable tool in psychiatry and neurology, and this book comes at a time when interest in brain stimulation in general, and in TMS in particular, is growing. In the United States, the FDA has cleared several devices to deliver TMS for the treatment of medication-resistant depression, migraine, and presurgical brain mapping, and international approval extends to treatment of developmental disorders, pain, stroke recovery, epilepsy, and dementia. Nearly one thousand TMS clinics exist in the United States, where patients who have seen little or no response to drug treatments and other more traditional interventions are being helped. However, knowledge of TMS and its potential remains limited in the medical community and among the general public, and many patients who could benefit from it are not being offered access. This is something that I hope will change in the near future, as stories like John's come to light. At the same time, more research into the effects of TMS on the brain are needed to improve its therapeutic potential and minimize its risks.

It is important to realize that John's story is not the experience of a patient undergoing a medical treatment. This was a research study that I led, approved by the ethics committee at Harvard Medical School and Beth Israel Deaconess Medical Center. I was not John's physician prescribing a treatment for autism. This study aimed to examine fundamental mechanisms of brain function in individuals with autism spectrum disorders and John was one of many study participants. As a study participant, John's experience is unique and personal. There is immeasurable value in his singularly subjective account, but the results that John experienced and that he describes in this book are his personal ones, and while they are incontrovertibly true and astonishing, they are not necessarily the objective outcomes of the study. That is important to consider. But that said, I do believe that John's results should inspire every interested science reader out there to wonder at the fascinating inner workings of the brain, and to be encouraged by the potential hope offered by techniques like TMS.

As a research scientist and as a physician who treats patients, my mindset and attitudes are often divided. As a physician, I know that my duty ultimately is to help my patient feel better. But as a scientist, I have been trained to transcend individual experience in order to learn fundamental truths about a disease, a brain process, a condition. If a research participant in one of my studies feels better, my task is to identify and understand the reason for the improvement. Often I am torn between using averages and statistics to blur the individual experiences of patients, and focusing on the experience of each study participant. This is a balancing act. And yet, as John movingly reminds us in his book, medical research involves people, and each patient can teach us invaluable insights.

Medical journals report on objective research findings, but sometimes, as in John's case, a participant's subjective experiences are unanticipated and surprising, and they potentially outweigh what we are able to capture with traditional measures of patient outcome. Listening to the experiences of a patient who participates in a study can be incredibly valuable, and I believe that John Robison's story begs a more patient-based approach to research.

The late Oliver Sacks was a masterful narrator of patient-centered medical histories, and in many ways, John's story reminds me of some of Oliver's most fascinating cases. But Oliver was primarily writing about others, whereas John's book tells a deeply personal tale. His astonishing story of transformation, of overcoming disability and deriving benefit from an experimental intervention that completely changed his life, is rife with inspiring lessons for each of us. It is a strikingly moving personal narrative about the nature of emotion, and about the opportunities afforded us when we seek to understand neurological difference.

ALVARO PASCUAL-LEONE, MD, PhD
Professor of Neurology, Associate Dean for
Clinical and Translational Research, Harvard Medical School
Chief, Division of Cognitive Neurology, Director of the
Berenson-Allen Center for Noninvasive
Brain Stimulation, Beth Israel Deaconess Medical Center

Prologue

THERE I WAS, doing seventy-five miles an hour in the left lane on the Massachusetts Turnpike. Suddenly, without any warning, I found myself transported back to a Boston nightclub, circa 1984. It was eight P.M. on April 15, 2008, when everything changed as I switched on my car stereo.

I was fifty years old, half deaf from hanging out with rock and roll bands in my youth, and tired from a long day working on cars. On top of that, I'd just left Boston's Beth Israel Hospital, where a team of Harvard neuroscientists had run an experiment on me, using high-powered magnetic fields in an attempt to rewire my brain and change my emotional intelligence. I've always been weak in that area because I have autism. Some autistic people have trouble talking or understanding language. Others—like me—generally talk fine and listen some of the time, but we often miss the unspoken cues—body language, tone of voice, and subtleties of expression—that make up such a big part of human conversation. I've always had a hard time with that. Luckily, my social disability is offset somewhat by my technical skills. But many of the gifts that help me make a living and take care of myself today also left me feeling lonely and broken as a kid. Some vestige of that hurt has remained in me, and that was why I had agreed to join the scientists on what several of my friends had called a crazy quest.

The idea of fixing myself with a fancy new treatment had sounded great in theory, but from what I had seen so far, it hadn't worked. The scientists had proposed using electromagnets to rearrange connections in my head. It had seemed like science fiction, and maybe that's all it ever would be. As I got into my car that evening after four hours at the hospital, I was more exhausted and annoyed than when I'd arrived. But otherwise, as far as I could tell, nothing had changed.

The drive to Boston had taken two hours and now I was facing another two hours to get home. *What was I doing there?* I asked myself. But I knew the answer—I had volunteered for this research study because the scientists had issued a call for autistic adults, and I wanted to "make myself better" in some ill-defined but powerfully felt way.

Those thoughts and a thousand others were all running through my head when I plugged in my iPod and music filled the car. I'd done that same thing a thousand times before and heard nothing more than songs on a car stereo. I hadn't *seen* anything at all—just the road ahead. This time the result was strikingly different. All of a sudden, I wasn't in my car. I wasn't even in my body. All my senses had gone back in time, and I stood backstage listening to the Tavares brothers singing soul music in a dark, smoky club.

Years ago I'd stood by those stages as the sound engineer, whose job was to make sure the machinery of the show kept running. These days I hung around the stage as a part-time photographer, following performers through my camera lens in the hope of catching that magic moment. This was something totally different. When I'd engineered rock and roll shows all I saw or heard were the little cues that told me everything was okay, or not. Now when I work as a photographer I concentrate so deeply on my subjects that I don't even hear the sounds of the show. That night in the car, the recorded music captured me and drew me into a world of a long-ago performance in a way I'd never experienced before.

The transition was instantaneous. One moment I was navigating traffic in my Range Rover and the next I was watching five singers in a nightclub.

Floodlights hung from the ceiling, illuminating the stage, and I stood just outside the lit area. To my left, on the stage, I saw the Tavares brothers in sport coats and bow ties, with a backup band on the side. A flute player stood in the background, whispering his contributions to the melody every few measures. Tavares is known to the world for singing "More Than a Woman" from the *Saturday Night Fever* soundtrack, but they had a long history in New England before that and a much larger repertoire of songs. Thirty years earlier I'd been a part of that world, working as a sound engineer and special effects designer. Many of the big Boston venues used my sound and lighting equipment, and I'd stood beside countless stages and watched more performances than I could remember. Was I reliving one of those now, or was this a figment of my imagination? I could not tell then, and I still don't know today. All I can say is that the experience felt incredibly real. I could almost smell the cigarette smoke on my clothes. And through it all, some separate part of my mind kept driving the car, though I only know that because I didn't crash.

Meanwhile, the sound of their voices was so clear that I let my mind run free. The musicians and their gear were right in front of me onstage. Looking into the wings I saw amplifiers and road cases stacked in the darkness. Scanning the club I saw the keyboard player, with his rack of instruments. One of the singers onstage walked toward me, and I heard the swish of the cable as he carried the microphone in his hand.

My vision was crystal clear, my head was full of sound, and I felt totally alive. The sterile digitized songs on my iPod had come to life and the feeling was so magnificently overwhelming that I began to cry. Not because I was happy or sad, but because it was all so intense.

I turned up the volume and sank deeper into the melody. The brothers kept singing, my car kept driving, and tears ran down my face. I felt the beauty of the sound wash over me, and every note was brilliant, new, and alive. This was similar to the way I heard music thirty years earlier, when I had spent every waking moment listening to performances, watching audio signals on my oscilloscope screen, or visualizing the sounds of instruments in my mind. Back then, "listening" was such a detailed experience that I'd recognize individual instruments and their positions on the stage. I'd hear the voices of each background singer, distinct, as he or she stepped up for a chorus. But now the experience was richer and deeper, with an added layer of feeling.

Suddenly I had an insight: Perhaps I was hearing music pure, and true, without the distorting lens of autism. Perhaps others heard that emotion all along, and now I could too. Maybe that was why I had cried—because I could *feel* the music, something that autistic people do not often experience in response to things we see and hear. I'd always been able to tell when music was happy or sad, but that night the Tavares brothers' music had hit me with a power that was new and unexpected.

A few hours earlier, back at the hospital, I had listened to two people shouting in anger as they passed in the hall. *He's mad,* I said to myself, without a trace of emotion attached to the observation. I was an accurate, logical observer. Now, as I listened to Tavares sing, tears ran down my face as I felt the emotions rise up from the lyrics of "She's Gone," "Words and Music," and "A Penny for Your Thoughts."

As many times as I'd heard those melodies sung, I'd never felt them the way I did now. Earlier that day, I would have understood the logical meaning of the words but nothing more.

At that moment, I got it. A song like "She's Gone" wasn't just words and melody, delivered to the audience with artistic precision. It was an expression of love, written and sung for a real person. I wondered who she was and what had become of her.

Later that night I sent a message to the scientist who was heading the effort. "That's some powerful mojo you have in there," I told him. And we were just beginning.

Switched On

An Electrifying Proposal

MY ADVENTURE HAD STARTED rather inauspiciously a few months earlier. I was standing by a table covered with cookies, at the entrance to the auditorium at the Elms College library in Chicopee, Massachusetts. The cookies were just standard school cafeteria fare, but someone had to eat them, so there I was.

I'd been invited to Elms to talk to students, faculty, and anyone else who might wander in on a cold January night. Elms had billed the evening's program as an "autism workshop," and I was its ostensible leader. That in itself was an extraordinary thing—me leading a college workshop. Until quite recently, the only workshop I'd ever run was the one at Robison Service, where we restored Mercedes, Jaguars, and Land Rovers. I wasn't a college professor—I hadn't even gone to college. I'd begun as a self-taught engineer who created sound and light effects for rock and roll shows. Twenty years ago I'd left that world behind and started a small business. Now I was a car mechanic with a side interest in freelance photography. However, I'd just written a book about living with Asperger's syndrome. (That was what clinicians called the type of autism I'd been diagnosed with. Today they call all forms of autism the "autism spectrum.")

I was already getting invitations to come talk about it in some pretty sur-prising places.

I'd grown up knowing I was different but having no idea why. The less obvious forms of autism—like Asperger's syndrome—were not widely recognized until the 1990s, and I wasn't diagnosed until age forty, in 1997. My discovery of how and why I was different was so empowering and liberating that I felt compelled to share the story with the world. The guys at the car shop thought I was crazy to take time off to write a book, but my brother, Augusten Burroughs, had written his own story, *Running with Scissors,* a few years before and I'd felt sure I could do it. Now my book was a reality, and its publication had connected me to more people than I'd ever imagined, all fascinated by autism.

First were the adults I'd met through my local autism society, a part of the Asperger's Association of New England (now called the Asperger/Autism Network, or AANE). They'd been great—a welcoming and sup-portive community that gathered twice a month to talk about the tribula-tions of everyday life. I was surprised by the extent to which autism tied us together, different as we seemed as individuals. And wherever I spoke I also met parents, many of whom seemed to take encouragement from the fact that I'd matured into an independent and self-supporting adult. Their reaction to my success made me think I'd dodged a bullet, living in ignorance of my diagnosis for so long. When I was growing up, I never for a moment doubted that I would be able to make a living. What other choice was there? Starvation? Yet many of the parents I was meeting seemed to doubt that their kids could do much more than get dressed in the morning and play videogames.

Their low level of expectation was shocking to me, and I began to won-der if it was an unintended downside to the new diagnostic awareness. Maybe today's autistic kids were like wise and wily pets who had trained their parents to feed them, house them, and provide entertainment and healthcare for a lifetime, all for free. When I offered this insight to a few of the mothers they did not find it amusing or enlightening.

I had done some speaking about autism before *Look Me in the Eye* was published—for schoolkids and even in jails—and I initially thought of the book as an extension of my in-person storytelling. After it was published I expected to reach a wider audience, but I never imagined the sort of response I encountered. I'd never gotten so many emails, calls, and messages, all from people with a stake in autism. I'd imagined creating a book as a cerebral, literary process, but no one seemed to be interested in the technical or creative aspects of my writing. It was all about autism. Everywhere I went, readers questioned me about the ideas I expressed and the things I believed. One of the first to make contact was Jim Mullen, then the president of Elms College. He'd gotten a prepublication copy, read it, and invited me to tour their campus and their new autism program. Jim introduced me to the faculty and asked if I wanted to get involved. They were developing a graduate program in autism therapy, and I was flattered to think that my ideas might make a useful addition to their curriculum. And I imagined it might be fun to surprise the faculty members who brought their Subarus and Volvos to my car complex for service. I'd always joked with them, saying, "You never know what a car mechanic will do next!"

That was my answer whenever a client of my car company encountered me moonlighting as a photographer on the edge of a concert stage, up on an acrobat's high wire, or in the circus ring with a lion. With my photographer's vest, ID lanyard, and three big cameras hanging off my shoulders, I sure looked different from the way I appeared in our service department. And being six foot four, I guess I'm a hard guy to miss. The funny thing was, I was perfectly in my element among performers and musicians, and they would have thought it just as strange to see me in the shop at Robison Service, surrounded by broken Jaguars and BMWs. Now publishing a book had led me to a new place—one where no one from my past would have expected to find me. The night of the workshop at Elms College, I'd had just four months to adapt to my new role as "autism expert."

Nowhere in my book had I claimed to be knowledgeable about anything but cars, electronics, and my own life. Nonetheless, readers said my lived autism insights made me an expert on that, and I was doing my best to meet their expectations.

The only worry that would sometimes creep into my mind was that I wasn't an accredited autism professional. My expertise was limited to my own experience of growing up different. The prospect of giving wrong advice by example worried me a lot, so I resolved to learn as much as I could about autism. I couldn't change the way people saw me, but I could change my foundation of knowledge, and I set out to do that as quickly as possible.

I never knew who would be in the audience at my talks. Some of the folks who came to hear me speak were established autism clinicians and therapists. I met teachers, counselors, psychologists, psychiatrists, and physicians. They often expressed fascination with my stories, and I wondered how to interpret that. Did they identify with my experiences? Or were they thinking something like, *This lab specimen can talk!* There was probably a bit of both.

Whenever a professional approached me, I listened very intently because I never knew when I might learn something vital. The problem was distinguishing genuine experts from trolls, opinionated laypeople, and the occasional crank. When Lindsay Oberman walked up to me that night at Elms College, I didn't know who she was or where she fit into the puzzle. She seemed the right age to be attending the school, and that's what I first assumed. She looked like a typical grad student—young, enthusiastic, and conservatively dressed in jeans and a sweater. Some of the people at the Elms event wore fancy jewelry or sported exotic tattoos or piercings, but Lindsay's only adornment was a handbag and a book.

Despite the simplicity of her appearance, she managed to stand out. Even now, I can't say what it was about her that made such an impression on me. Maybe someone who reads people better than I do could answer that, but it was enough for me that I sensed she was smart and different.

"I'm a postdoctoral researcher from Beth Israel Hospital," she said as she introduced herself with a business card that read "Dr. Lindsay Oberman, Ph.D." "We're doing some autism studies and I'm hoping you'll let me leave some flyers about our research. We need adult volunteers for a project we're starting in the area of improving emotional intelligence for people with autism."

Now that was a new one.

I pictured the audience for my talk as fish in a pond, and Lindsay on the bank with a fishing pole and a net, scooping the ones she could catch into a bucket and carrying them away to some unknown fate. I wondered what she intended to do with these adult volunteers, imagining psychological tests followed by stew pots for the losers.

That unsettling vision left me unsure how to respond. Was she asking me to endorse her research and encourage people to volunteer? I didn't even know what her study was about. So I asked her.

She began to describe her interest in autism and her desire to remediate some of its disabling symptoms. "We're experimenting with a new technique called TMS, which stands for transcranial magnetic stimulation. We use an electromagnetic field to induce signals in the outer layer of the brain. We're hoping to develop a therapy that helps autistic people read emotion in other people."

That last line got my attention. I almost said, "That's exactly my problem," but I kept my mouth shut. My grandfather had taught me never to show interest in something that was offered for sale. It only makes the price go up. Even though she hadn't said a word about money yet, for all I knew, she was going to end her presentation with the news that I could sign up—today only—for a special introductory price of $1,999.

But she didn't mention money at all. Instead, she launched into a five-minute explanation of mirror neurons, electromagnets, and pulse energy. I wasn't sure if Lindsay had read my book or knew about my background as an electrical engineer. What she described sounded very similar to the lasers and sound systems I'd worked with eighteen years earlier. The difference was, our electromagnets were part of loudspeaker arrays that filled arenas with sound, and our pulse lasers scattered pinpoints of light over crowded dance floors or bounced signals off the moon.

She proposed using similar technologies on people, by pulsing electromagnets to fire microscopic shots of energy into the brain. I'd never considered such a possibility, but I was intrigued.

And the mirror neuron thing was fascinating too. I'd recently read up on mirror neurons—brain cells that cause us to act out what we see or hear. We see our mother smile at us, and our mirror neurons make us smile in response, sort of a monkey see, monkey do effect (literally so, because it was first observed in monkeys).

The idea of stimulating mirror neurons with electricity sounded more than cool to a techno geek like me. I had a brief vision of Frankenstein's monster with lightning sizzling between his ears, but I understood that this would be something far subtler. Years ago, we'd fired thousands of watts into lasers and loudspeakers, but the brain operated at power levels a million times smaller. Delivering tiny pulses of energy to alter the process of thought sounded like a fascinating challenge. I'd have jumped at the chance to design their equipment if I still worked as an engineer.

Lindsay had captured my attention right away with her talk of medical magnets. Maybe it was the use of familiar technologies in a completely unexpected way, or perhaps it was the hope of unraveling my social disability through applied electrical engineering—either way I was hooked just as surely as a guy in the funnies who sees a pretty girl and gets hit by a thunderbolt.

Was it possible to use energy to change the brain? It sounded like science fiction. "It's definitely science fact," Lindsay assured me. "When TMS adds electromagnetic energy to the neural networks inside your head, it helps them build new connections, and it reinforces the connections we want to strengthen.

"I've worked with it in the lab," she told me, "and I've even had it done to me, so I know it's safe." Until she said that I hadn't even stopped to consider whether jolting the brain with energy might be dangerous.

I fired questions at her as fast as I could think them up, and she was eager to answer, tossing out terms like "cerebral cortex" and "brain plasticity." But when I asked about power levels, polarity, and patterns of electrical waves, I discovered that her knowledge of the physics of TMS was limited. Lindsay was a user of electromagnetic technology, not a maker of it. Her training was in neuroscience as opposed to electronic circuit design. She was quite fluent in the language of the brain, but from my perspective, as a newcomer to neuroscience, hearing which cerebral areas she might stimulate didn't mean much because I didn't know one from the other. And Lindsay wasn't familiar with any of my electronic engineering terms; all she could tell me was that different patterns produced different effects. I asked what the effects were and she mentioned two terms—"potentiation" and "depression." When I asked what those terms meant she explained that they referred to energizing or turning down particular areas. "If we depress your speech center, you'll have a hard time talking," she offered as a quick illustration.

When I asked her exactly how that happened, she couldn't answer me. I wasn't sure if she didn't know herself or if the answer was unknown to science. Either way, I wanted to learn more.

"My boss can explain how it works better than I can," she said. She wrote down his name, Dr. Alvaro Pascual-Leone, on the back of her business card and invited me to meet him the following week.

The idea of using pulsed magnets to change the brain was fascinating to me for another reason as well. My family had a history of mental illness, and I'd always hoped for some kind of breakthrough. My mother had experienced semiannual psychotic breaks throughout most of my teen and adult life—I'd seen her sent to the state hospital and tranquilized into a zombie-like stupor—until her brain was reconfigured by a stroke when I was thirty-three. Her doctor made a remarkable observation two years later, when she was in a rehab hospital, paralyzed on one side and having lost much of her speech. "The stroke seems to have killed the part of your mom's brain that made her become psychotic. As hard as this is for all of you, that is an unexpected silver lining."

To make matters worse, one of her brothers was schizophrenic, and her father suffered all his life from serious depression. Given my family history, I always wondered when the other shoe was going to drop for me. Was there a way to fix the broken things inside our heads while leaving the remainder untouched? Could we alter speech, coordination, or vision while leaving personality intact? Might we even adjust personality?

Lindsay's description of TMS hinted at an answer. But she didn't promise any benefit at all, particularly for me. "This is a research study," she told me. "It's not a treatment trial. That would come later, if we find something that seems to work." But even without any guarantees, I was ready to sign up. I'd had a lifetime of feeling I was less than everyone else. After fifty years I'd come to accept my lot in life, but now that I saw a chance to leave second-class citizenship behind I was going to grab it.

Then I had a crazy thought: What if her boss was an engineer? Nothing could be finer! We would solve the problems of autism as fellow engineers and use medical magnets to create autistic supermen! We'd win fame and glory! Alas, my hopes were immediately dashed. "He's a neuroscientist," she told me.

When I first started giving talks about autism, I had encountered some unusual theories about my condition. "Don't you realize that your autism is caused by mercury poisoning?" I was asked more than once. One determined mom insisted that dangerous chelation therapy was the key to a healthy, autism-free life, just as surely as Brother Love's Traveling Salvation Show offered the road to redemption. Another well-meaning parent professed the wonders of hyperbaric chambers. If chelation didn't fix me, that surely would. Prior to meeting these parents, I'd never thought of myself as being in need of a cure. The idea that my autism was a disease or a vaccine byproduct had always felt somewhat insulting. I did not know why I was different, but I knew in my bones that I was not the twisted spawn of some pharma-government vaccine conspiracy. Nor was I a different kind of human, escaped from the Alien Containment Unit in Area 51. To my great relief, Lindsay didn't suggest any of those things. She didn't suggest much at all about how I came to be autistic. She just proposed a possible way to help my brain rewire itself to work a little better. With the good bit I knew about electromagnetics and the nothing I knew about brain science, that put her suggestion a thousand miles ahead of anything I had heard before.

And when she was done—after my earlier worries about cost—Lindsay told me I'd receive the princely sum of fifty dollars for every session. What a deal!

TMS was the first therapy I'd heard about that made sense to my engineer mind. The idea that electromagnetic coupling could deliver controlled energy to small parts of the brain appealed to me because I knew it was possible and I had always been dubious of psychiatric drugs. How many billions of unaffected cells did they touch and change? To me, taking a psychiatric drug was like pouring oil all over a car when the low oil warning light came on. Doing that might get some oil into the engine, but it mostly just makes a mess. Drugs in the bloodstream work the same way, diffusing through the whole body. TMS, on the other hand, targets a tiny focused area. It didn't take a medical degree to appreciate that difference.

Later that night I looked up Lindsay and her boss online. That was when I learned that Beth Israel is a teaching hospital of Harvard Medical School, and that Pascual-Leone, MD, PhD, was both a medical doctor and a neuroscientist, a full professor at those august institutions. Lindsay had introduced herself by her first name, but reading about her I now wondered if I should have called her Dr. Oberman and whether I'd addressed her with enough respect back at the auditorium. She'd hardly looked older than my son, Cubby, who was just out of high school. But there she was on the website, with a doctorate from the University of California, San Diego, and a faculty appointment at one of the most prestigious universities in the world. I reminded myself that I was not a great judge of age and that appearances could be deceptive.

Lindsay had told me she had done her doctoral work with V. S. Ramachandran. As it happened, I'd just finished reading about Rama and his groundbreaking work with phantom limbs in a book called *The Brain That Changes Itself,* by Dr. Norman Doidge. Lindsay's former professor was a legend in the field of cognitive neuroscience, and I was duly impressed.

Rama had a fascination with autism too, and Lindsay had said she discovered her own interest while studying in his lab. When I considered her lack of electrical engineering knowledge, I reminded myself that you don't have to know how the hardware in a computer works to be a star programmer. The next time I saw her, I made a point to ask her what she thought of the comparison. "I'm not sure anyone in the field truly knows how brain circuits work," she said, "at least not at the level of a computer chip." I would soon discover how incredibly complex the brain is, orders of magnitude more intricate than any circuit.

Yet I couldn't help trying to relate what she had said to my own experience with electronics. When I worked as an engineer in rock and roll, I created custom instruments but I never learned how to play them except in the most rudimentary way. The fact that I could create a custom guitar without much musical knowledge and then a musician could pick it up and make beautiful melodies with no idea how it worked inside had always fascinated me. Perhaps Lindsay was like a musician of the mind.

That reflection led me to an unsettling thought: if she was the musician and I signed up for her study, that would make me the instrument! I remembered all those nights at concerts watching rock and rollers hammer their guitars till the strings came off, and I hoped that wasn't what was in store for me. When a musician gets a hit record, his guitar doesn't generally jump for joy.

Still, the conversation had gotten my hopes up. TMS sounded like a doorway to a fascinating new world, one I very much wanted to enter. I just hoped my natural rudeness hadn't chased Lindsay away.

The Value of Detachment, Circa 1978

A PAIR OF LEGS stuck out from beneath an old Ford Torino on the lower road from West Springfield to Holyoke. The guy they belonged to was clearly dead because the legs weren't moving, and when I walked closer, the car on top of him was still and quiet, its engine cooled. He must have been there awhile. There was another guy in the road about six feet away, sitting with his legs crossed, slowly rocking back and forth, not saying anything. When I called out to him he just ignored me. I wasn't sure what his connection was to the scene—whether he'd been in the car or what. He didn't look injured, and first aid wasn't going to help the guy under the car. It was about 4:30 on a Sunday morning when I came upon them, driving home after a long night's work.

I'd come around a gentle curve and seen the car and a person in the road. From a hundred yards away, it looked like a wreck with someone maybe thrown from the vehicle. My workday ended well after the bars closed, and I seemed to pass drunk driving carnage almost every night on my way home. When people crashed before last call, the cops and the ambulances tended to show up immediately. But the roads were empty by three A.M., and if you wrecked in the hours before dawn it could be a while before anyone found you.

Today you just dial 911 on a cellphone, but this was the 1970s, and cellphones didn't exist. There weren't as many people on the roads either.

I looked back at my car to make sure it was secure. Sometimes you stopped for a crash or a breakdown and some lowlife hopped in your rig and sped away. There was no one else in sight, which was just as well by me.

I'd never known what to say to people in casual conversation, but that didn't matter at accident scenes and emergencies. In these situations, I always knew what to do—I followed the rules I had learned as a boy. My great-grandfather had been the county agent for the U.S. Department of Agriculture in Gwinnett County, Georgia, and his first cousin was the sheriff in nearby Carroll County. They never succeeded in teaching me any manners, but I had always been pretty good around machines, and they taught me what to do in a crisis.

The first thing I did was stop my car right in the center of the road. I pointed my low beams at the Torino and turned on the hazard flashers. Lots of drivers on the road at that time of the night, drunk or asleep at the wheel, might have run that second guy right over. I was sober and alert, and my car would protect him.

Sometimes the wrecks I encountered were bloody and noisy. Not this one. It was strangely clean and silent. The guy was pinned under the car from his waist up, but there was no blood. He wasn't breathing or making a sound. Of course, you wouldn't be breathing either, with three thousand pounds of car on your chest. It was that time of the night when most of the drunks were already home in bed. The crickets were done chirping, and the birds weren't ready to herald sunrise. That was at least an hour away.

The only light on the scene came from the headlamps on my car, which I'd left running to keep the battery charged. Street lighting hadn't yet come to this part of town, and there was no telling how long I would be there. Once you stop, you're committed.

"What's going on?" I asked the guy on the ground a second time, but he still didn't answer, just rocked back and forth. I'd seen some people act that way when they were high on drugs, and if you touched them they blew up, all crazy and violent. I'd seen other people look like that because they were in shock. This guy didn't seem to want anything from me, so I let him be and looked around.

Today I realize that most people would be horrified by what I was seeing, but I was not affected that way. That's one way autism has shaped me and set me apart from the majority of humanity, even though I didn't know it then. The sadness or horror of the accident simply didn't register with me. I didn't pick up on emotional cues from other people, and the two men at this crash scene were both strangers to me. *Why should their plight mean anything to me?* That's what I'd have said at the time, but I still acted to secure the situation and protect the survivor in the road. Even without feeling the expected emotions, I did the right thing.

Being emotionally blind isn't the same thing as being uncaring or amoral. My sense of right and wrong was quite well developed, and I did the best I could for other people. It's just that my senses and abilities were limited, so I didn't always do what they expected.

A brief flash of fear washed over me, but I quickly realized there was no threat. A walk around the car told the story. The left front tire lay on its side just outside the wheel well. At first I thought it had been torn from the car in a crash, but I quickly realized that was wrong. A rusty old bumper jack lay where it had fallen, on the ground next to the front bumper. This guy hadn't had a wreck at all. He'd just had a flat tire. But for some reason, he crawled under the car, the jack slipped, and that had proved fatal.

"Never climb under a car unless it's on a safe stand," my grandfather said the first time we changed a tire, and I never forgot his words. Turning back to the guy in the street, I said, "You okay?" There was still no answer. I looked a little closer at him and saw no sign of injury. No need to risk moving him and maybe starting a fight. There weren't any pipes dripping fuel or wiring making sparks. The scene was stable, and that was all I could ask.

I stood there motionless, breathing the crisp night air and pondering my next move.

Pop! The tinny bang of metal on metal might not have been very loud, but in the predawn silence, it sounded like a gunshot. I jumped a foot into the air and spun around to face the noise. A slight rattle showed me the source—a lug nut that had been sitting on the hub had fallen off and landed in the hubcap that lay on the ground by the front bumper. As I watched it roll back and forth in the dish of the hubcap, I wondered what other things might be slithering around unseen in the dark.

Stuff settles, I told myself, but I still felt a creepy chill, and the hairs on the back of my neck bristled. The darkness felt thick, like it was closing in, and I knew it was time to do something.

But what? For a brief moment I considered leaving, but I was wary of getting arrested for driving away. Better call the police. I walked down the road to the nearest house, about a hundred yards away. No one answered the door of that house or the next one. A grizzled old man with a bath-robe, a baseball bat, and a strong smell of liquor opened the third door.

The denizens of lower Holyoke are known far and wide for their pug-nacity, and this one was no exception. Luckily he was slow moving in his freshly awakened state. "Call the cops," I told him. "There's a wreck out there, and someone's dead." He looked at me, looked down the road, and shut the door in my face without a word.

A more assertive man might have kicked the door down at that point and repeated the request. But I'd looked him in the eye—like my grand-daddy had told me to do—and now I trusted him to make the phone call. Anyway, I'd banged on enough doors. I walked back to my car and sat down to wait, hopeful that he'd do what I'd asked. Ten minutes passed, and my confidence was starting to waver. Then I saw the blue lights com-ing. Climbing out of my car, I stepped to the side of the road. Cops are skittish late at night. You have to move slow and be out in plain sight. Criminals run. Law-abiding citizens and people with good political con-nections stand their ground. As soon as the cruiser stopped, I raised my arms to signal a greeting and show I didn't have a gun. There was just one officer in the car, and he got out and approached me warily. I explained the situation.

The cop walked over to inspect the scene. He reached down to touch one of the dead guy's legs, perhaps to see if it was hooked to a real person. To me the answer had been pretty obvious, and my cousin the sheriff had always told me never to touch anything that might be part of a crime scene. The cop stood up quickly and walked over to the guy sitting in the road. He didn't get any more out of him than I had. "Wait here," he told me, and went back to his cruiser. There were some barks of voice and static from the radio, and a few minutes later police cars arrived from both directions. An ambulance showed up, and the attendants scooped the live guy out of the road and drove away. Another ambulance arrived and sat there, waiting for a wrecker to lift the Torino off his friend.

No one paid any attention to me, but I knew better than to drive away. Right then I was a witness, but if I snuck off I'd become a suspect, which was a lot worse. It was pretty clear to me that the jack had fallen and the guy had been crushed, but how that came to be remained a mystery. Maybe the other guy had been jacking it while his buddy was underneath. Why was he under the car anyway? You don't have to crawl under a car to change a tire. It was all just speculation for me, but it was an investigation for the cops.

Half an hour passed, and a part of me was thinking I should never have stopped in the first place. The affair wasn't any of my business. But sometimes a person who's able to help makes all the difference. I'd wrecked my motorcycle a few years before, and I'd have been in a bad way if strangers hadn't stopped, blocked the road, and kept me from getting run over by the oncoming traffic.

The first cop on the scene came over to question me with another, more senior officer. I told my story—what little story there was—and showed them my license. Then I mentioned the tire and the jack and what I figured had happened. "Did you touch or move anything?" the older cop asked me. "No," I said, "I didn't touch a thing. I tried to talk to the other guy who was sitting there, but he wouldn't answer me. He never said a word or moved from the moment my headlights hit him to when you guys took him away."

The other cop was writing it all down.

"What were you doing out here?" We were at the edge of a rough part of town, and the older one must have thought I looked suspicious. "Going home," I answered. "I install sound and lighting systems in clubs and I work late some nights. Tonight I was at the Arabian Nights, setting up a new illuminated dance floor." The cop looked at my car—a blue Cadillac Eldorado convertible—but he didn't say anything. He could tell I hadn't been drinking, and I wasn't acting aggressive.

Many of the city's biggest nightclubs were owned by connected guys, or guys in the rackets. There was a lot of money in a popular place, and all of it in cash. But I wasn't in the mob and I didn't own a club. I just worked in one.

And I didn't live in their town. My destination lay across the river, in South Hadley. In Holyoke a car like mine might signal a pimp or a drug dealer. In my town, Cadillacs belonged to insurance agents or doctors. I probably didn't seem like either of those things. Mostly, I was just tired, and I'm sure I looked it. "You got a phone number in case we need to call you?" I gave it to him, but I never heard from those cops again.

That's what life was like for me as a young adult. People made fun of my stilted manner, my pedantic speech, and my detachment from other people. I didn't see anything wrong myself—I was just logical. But the people who knew me seldom saw that as a gift. That made me sad, but my logical disposition still protected me when I stumbled into bad situations. And that happened a lot, working in bars and concert arenas. At a time of the morning when most guys my age were asleep in their college dorms I was working, by myself, in a nightclub. In those places I was often surrounded by pretty girls, good liquor, and great drugs. I'd have partaken of all of that and more if only I'd known how. But most evenings all I went home with was broken sound equipment, and my company that night was a Yamaha power amplifier. I had no clue how to chat someone up at the bar, even as I watched others do it with what looked like no effort at all.

It's funny how that turned out. I'd always known I was different. Kids used to tease me about my special interests and fixations. Now musicians and nightclub owners paid for my insights into sound and lighting. Unfortunately, that only went so far. Technical prowess had bought the Caddy and nice clothes, but it hadn't won me many friends.

Why was I that way? At the time, I didn't know. Some people called me robotic; others said I was disconnected or uncaring. Why did I remain cool as a cucumber at the kind of accident scenes that would have rookie cops staring in horror?

All I knew was that I was a good guy in a crisis because my logical mind took charge. Today I know that it's because I am autistic, and autism has given me a mixture of disabilities and gifts. Being blind to the emotional cues of other people can be crippling, but my sense of logic and order has also been a great benefit.

I walked through the scenes of my life like an outside observer, stepping carefully over the rubble and staying out of trouble. There was very little happiness in my world. Luckily I had a natural gift for understanding machines and making things work. But people were a complete mystery to me. I wanted desperately to have friends and be popular, but the closest I could come was working with them to do a job. Even there, I was more of a lone wolf.

That kept me on the fringes of society. The few friends I had were fellow freaks and misfits. The people I worked for in those years were mobster club owners and crazy musicians. My dream was to have a real job, one where I put on nice clothes and worked in a clean office, but I was light-years away from anything like that.

I wasn't a popular kid growing up either. There were times I didn't have any friends at all. Other times, I had a few. My childhood was mostly lonely, but maybe that was just as well. Looking back, I think my oblivious nature protected me. Other kids called me names and insulted me, but most of it rolled right off. I heard their words clearly but the true meanings escaped me. Today, when I recall some of the bad things people said, I realize (with the benefit of adult hindsight) that there must have been a million other almost-as-bad things that didn't penetrate my awareness. My only joy came from reading and building things. I had Tinkertoys, Lincoln Logs, and finally a real Erector set. That was where I found my first success. I may have been a social failure, but I was an extraordinary child engineer—at least in my own mind.

Unfortunately that hadn't translated to academic success. Teachers said I was smart but lazy. I might not have been into my schoolwork, but I got better and better at understanding machines, and then I discovered electronics. That was my ticket to success. At sixteen I left home and got a job. School wasn't taking me anywhere, and my home life had become pretty ugly. My mother was descending into mental illness and was briefly institutionalized when I was a teenager. Meanwhile, my father drank and became violent most nights. The combination was enough to drive any-one away.

Some people remember a girl from their childhood. For me, it was a Massey Ferguson 135 tractor that taught me the secrets to making my way in life. Engines and machines were my first great love, and they were what saved me. I've come to realize that my aptitude for machines is pretty unique. The average person stands next to a running motorcycle and hears nothing but noise. Their main interest is to shut the thing off, so it will be quiet. Not me. I hear the individual pops of the combustion gases and all the other noises from the engine as a mechanical symphony.

The joy I got from communing with engines pointed me toward my next great interest—electronics and music. By the time I was in my twen-ties I had found a home in the music business, where I made a good living doing weird technical stuff no one else could do. My first gigs were with local bands, who gave me an entrée to working bigger and bigger shows. By the mid-1970s I engineered a lot of the big productions at the Univer-sity of Massachusetts, where my parents taught. That was where I met the crew from Britannia Row Audio, fresh off the boat from London. The guys at Pink Floyd's sound company hired me to fix a mountain of broken gear they had shipped to America. When I did that successfully, they gave me the chance to create some really kickass special effects. A few years later, I went on to design the fire-breathing and rocket-launching guitars that became a KISS trademark. My engineering skills made me a lot of money, but I was still alone, and I was keenly aware of that fact. That put me in a strange position. My technical accomplishments were fairly visi-ble, and countless young people dreamed of having a job like mine—building gear for famous musicians and seeing my stuff play in the big concert venues like Madison Square Garden. Folks would say things like

"Boy, I wish I could be you!" I thought they were nuts, because none of that mattered to me—I'd have traded all that I had to feel popular and loved.

That was the hidden disability of autism in me, even though I didn't know it at the time. The result: I quit the music business thinking I was a failure, because I couldn't tell how people felt about me and my creations. There had been a sense of kinship living with the musicians in my first band. That vanished when I started to work with the bigger, more glamorous groups. When I was on tour, I was either working or alone in my room. When I wasn't touring, I was by myself in my lab designing and building gear. The community of musicians and artists that had welcomed me as a teenager seemed to have left me behind in my twenties.

So I took a corporate job for a while, engineering electronic toys and games for Milton Bradley, and then later I worked as a consultant for government contractors and running the power systems engineering group for a laser manufacturer. The technical work was fun, but the interpersonal parts of the jobs were not. People told me I said unexpected and disturbing things, and my words got me into trouble more than once. "Good engineer" was how I wanted to be described, but "jerk" was the more common appellation, even as my circuit designs were superstars. My bosses told me I was quick to dismiss incompetence in others and was not much of a team player. As I advanced in the corporate world, it was increasingly evident that the people who made it to the top were the total opposite of me. People skills got you to the pinnacle, and it seemed like emotional cunning helped keep you there. The ability to say one thing and feel another was a good trait for a manager, as was the ability to dissemble smoothly on demand. Neither of those things was me.

After two layoffs and more than one threat of dismissal I decided to quit on my own terms. I left electronics and went back to cars. They certainly felt safer; I could work for myself and there wasn't anyone to fire me. There weren't any office politics to navigate either—just me and my four-wheeled friends. I had to answer to my customers, of course, but if one got mad and left, the others would remain. That seemed a lot less risky than having one boss I could never figure out.

I started fixing old Mercedes and Land Rovers in my driveway, and the business grew and prospered enough for me to move it to Springfield, to a bigger commercial garage. The flow of work increased, and I hired mechanics to assist me. Suddenly, I was the boss!

Just before turning twenty-five, I'd married my high school sweetheart, Mary Lee Trompke. I called her Little Bear because she was short, belligerent, and my best and sometimes only friend. We fought quite a bit, but we had a deep bond from our shared dysfunctional childhoods, and at least initially that was enough to sustain us through the bad times. A few years later we had a son, named Jack in honor of my grandfather, but to me he was always Bear Cub or just Cubby.

There was nothing better than watching our little boy grow and thrive, and I told everyone how much better he was than me. That was true, except on the playground or in school. Seeing my son commit one sandbox gaffe after another—and alienate the tykes around him—brought back painful memories of my own childhood. I did my best to coach him, using my own experience as a guide. It seemed to work, because he started making friends in grade school—something I never really mastered.

Schoolwork was another matter. Cubby struggled mightily to read and master his assignments. His mom and I tried to help, but we didn't get too far. It was clear that there was a problem, but we had no idea what it was.

That was where things stood when I learned about my own Asperger's and its place on the autism spectrum from a therapist who also had become a client of my car business. "It's a new thing they're talking about in the mental health world, and you could be the poster boy for it." He left me with a book—*Asperger's Syndrome,* by Tony Attwood, an Australian psychologist. Once I got over the shock of his announcement, I started reading. Ten pages in, I knew in my heart he was right. Whatever this Asperger thing was, I had it.

Looking at my son, I saw the same signs in him. But I still believed he was better than I was, and in the first flush of awareness, I felt sure Asperger's was not a good thing for him to have. *He's not like me in that way,* I insisted to myself, even though I knew deep down that he was. A decade would pass before he got his official diagnosis, but the differences in him were there all along. Just as they were in me.

I occasionally encounter people today who tell me they wouldn't be able to face an autism diagnosis in their child. Their words take me back to the days when I learned about my own diagnosis but denied the obvious about my son. Today I say my diagnosis was a huge relief and a source of empowerment, but it took me a while to come to that realization.

I kept reading and schooled myself in the history of Asperger's and autism. Some doctors actually called my condition the "Little Professor syndrome," because kids with Asperger's can speak with great precision and detail about the things that interest them, often becoming fascinated to the point of obsession. Reading those words I was struck by how accurately they described me, and I didn't know what to feel.

Should I be happy? Should I be sad? Should I be relieved? I felt all those things and more. Learning about autism and Asperger's was a huge deal for me, because it gave me a legitimate explanation for why I was different. Hearing that I was different and not defective was certainly inspiring, though it took years to fully sink in. Reading about how folks with Asperger's typically act helped me to see some of my own behaviors in a new light. With that insight it became clear why some people reacted negatively to things I said or did.

Having a neurological explanation does not change the fact that I often acted like a jerk. That's one of the problems with growing more self-aware—you have a lot more moments where you think back on things you said and did and cringe in shame.

By the early 1990s I was spending all my time in Springfield, trying to make my car business a success. But the ups and downs of commerce at times shook my business, and my confidence. Repairing cars had seemed simple, and I was pretty good at it, but mechanical skill alone wasn't enough. A successful auto service company was as much about fixing the motorists as their vehicles, and that was something I was slow to understand. Meanwhile, Little Bear was increasingly immersed in a world of science fiction and graduate school. At night I'd find her reading fantasy and sci-fi till the wee hours, and her days were filled with the study of Central American culture as she pursued a doctorate in anthropology. It began to seem that we had very little in common outside of our son, and we ended up getting divorced as Cubby entered third grade. Miserable as my marriage had been at the end, its failure shook me something terrible. Getting married had seemed like a big step toward being a normal adult. Having that unravel made me feel like a total misfit and failure. Whenever I started to feel good, it seemed that something would happen to knock me back.

Dr. Attwood's book about Asperger's sustained me through all of that. Reading it backward and forward till the pages fell out, I resolved to make myself "normal." As an autistic person, there were a lot of things I missed—like the nonverbal signals other people sent one another. Learning what I was missing from a book didn't help me receive the signals, but knowing that they were there was enough to change my behavior, and I taught myself to emulate the social behaviors that others seemed to pick up instinctively. My gradual successes made me feel better about myself, but my growing confidence was always tempered by a trace of sadness, because now I knew: this is how I am and my behavior is immutable. I might learn to act more appropriately and fit in better, but I would always be different and socially, well, dumb as a rock.

It took a few years, but things finally started looking up again. I'd gotten a new girlfriend and we began spending more and more time together. Martha was the opposite of my first wife—quiet where Little Bear had been loud, neat where she had been messy. She had also grown up essentially self-employed, working in her father's business. Now she lent a welcome hand, building a presence for Robison Service on the newly ubiquitous Internet. Luckily she and my son got along, and life settled into a new routine as we moved in together. Cubby split his time between our place and his mom's.

In 2003 Martha and I got married, and we built a new home in Amherst so Cubby could attend the same local high school that I'd attended. "It's one of the top-rated schools in the state," people told me. Somehow I put aside my own memory of their tossing me out of ninth grade thirty years before. *Cubby is better than me,* I repeated, *and he's going to succeed where I failed.*

Meanwhile, the business prospered. I'd picked up enough interpersonal skills to keep some customers coming back, and the company grew steadily. At the same time, thanks to the tips in Dr. Attwood's book, I began making more friends. I wasn't the life of the party by any means, but the changes I'd made in my behavior—responding to people, saying the right things, acting the way others expected—made me a thousand times more popular than I'd been as a kid. I attributed much of the good that happened to my new self-knowledge and what I did with it. So much that had been mysterious finally made sense. Now people seemed to assume I was just eccentric, whereas before they'd thought I seemed crazy, or worse.

I felt like it was time to give something back, and I began visiting places like Brightside, a group home for kids who'd been taken out of bad or dangerous family situations. Speaking to teenagers who were "different" seemed useful for them and felt fulfilling for me. Everywhere I went, I got the same message. People want hope. They want to see people who come from bad—however that's defined—and end up making good. I realized my story might well be useful to young people who were like me and trying to find their way but not knowing how. That was what made me decide to write a book, and that changed everything.

Learning about autism, and then sharing my own story with the world, had been an important step in my life. Perhaps this TMS adventure was going to be the next one.

Medical Magnets

"IMAGINE YOUR BRAIN as an electrical organ. We propose to rebalance it by adding tiny amounts of electricity." That was what Dr. Pascual-Leone said to me across the dinner table the first time we met. Lindsay had suggested getting together when she'd introduced herself at Elms College a few weeks before, and they had agreed to join me at my favorite Boston area restaurant.

Alvaro was about my age, medium-sized, and was dressed in a tweed sport coat like a college professor. Which, actually, he was, in addition to being a practicing neurologist and a research neuroscientist.

My friend Richard had cautioned me about neuroscientists. "They're the most cold and inhuman of doctors," he warned. "All they want is to experiment on your mind, and they won't care what happens to you later." I'd raised my eyebrows a bit at that warning. Unspoken messages—like *I'm bored,* or *I'm excited,* or *Let's get out of here*—go right over my head, yet in spite of my social oblivion I've always had a decent sense of self-preservation. My ability to sense danger and recognize dangerous people has always been very good indeed, and nothing in this situation had set off my radar. It made me wonder whether Richard's opinion was based on real life or old horror movies.

Alvaro had struck me as warm and friendly from the moment we met. He told me about having a relative whose autism was more disabling than my own, and even with my limited ability to read people, it was clear that he was a caring man with a personal stake in the research he was doing. I trusted him right away.

Alvaro had come to Harvard from Spain and he was certainly an expert in his field, with both MD and PhD degrees and twenty years of experience at some of the top university hospitals in the world. I would have thought that a guy of his stature would be a bit standoffish and would want to be addressed formally as Dr. Pascual-Leone, but he was friendly and approachable from the moment we met. He'd even agreed to come to Legal Sea Foods! (Or maybe he simply liked boiled crustaceans as much as I did.) Despite his impressive credentials he was just Alvaro to me, just as Dr. Oberman was simply Lindsay. They took turns answering my questions, with Alvaro taking the lead. He spoke articulately and with a bit of an accent, enough to make me listen closely.

Between mouthfuls of swordfish and shrimp, he described TMS and how it worked, how he'd gotten involved with it, and how it might affect my blindness to other people's unspoken emotional cues.

The brain is an electrical matrix, he explained, with each brain cell or neuron connected to countless other neurons through a maze of microscopic cables and junctions. Someone familiar with electronics might equate the circuit board traces and gates in a computer with the axons, dendrites, and synapses of the brain. But the circuitry of the brain is infinitely more complex. The biggest electronic chips we can make have just a few million transistors on board, whereas a human brain contains about 86 billion neurons, each with hundreds or even thousands of connections. Counting all those connections would be like counting stars in the sky . . . they just go on forever.

To get a sense of that scale, consider one of the smallest things you can see—a grain of salt. A typical grain is about four thousandths of an inch in diameter. Two hundred and fifty of them, lined up on the edge of a ruler, would only stretch an inch. The average neuron is one-tenth that size, which means your brain packs at least a million into one single square inch. Stacked in three dimensions, one cubic inch of brain may hold a billion neurons—each capable of thought at some level. And that third dimension is what sets human brains apart from computers. Computer chips are essentially two-dimensional devices in three-dimensional packages. Most of the space inside a computer is empty air. The space inside your head, however, is filled with brain cells in numbers that defy comprehension.

Axons and dendrites are the electrically conductive threads that connect the neurons together, joined at points called synapses. This living wiring inside our brains carries thoughts from one area to another. Even as they do that, the biological wires in our brains can also pick up outside electrical energy through the process of electromagnetic induction. You might remember induction from science class, where the teacher waved a magnet next to a coil of wire, and the motion of the magnet induced enough energy in the wire to move a meter's needle as the class watched. Our brains have thousands of miles of microscopic wire strung between their neurons, all with the potential to receive electromagnetic radiation.

"By pulsing a powerful electromagnet next to your head, we can do the same thing inside your brain," Alvaro explained. "We can induce tiny electrical currents in those microscopic threads. Depending on their polarity and pattern, the currents we induce could enhance or inhibit the function of neurons we affect. The more tightly we focus the magnetic field, the more concentrated the effect."

His explanation made me remember that I had heard of TMS before, in a 2003 *New York Times* story called "Savant for a Day." In the piece, writer Lawrence Osborne described becoming a better artist after experiencing TMS at Dr. Allan Snyder's lab in Sydney, Australia. He walked into the lab with nothing more than the ability to draw primitive stick figure cats, and after a short session of TMS he was rendering the creatures like a virtuoso. He'd written, "Somehow over the course of a very few minutes, and with no additional instruction, I had gone from an incompetent draftsman to a very impressive artist of the feline form."

And then his talent had faded away, leaving him as he'd been before. Who wouldn't want to try something like that? A few other writers have since described their own experiences with TMS—how it can alter speech, sight, creativity, and even mood—but I hadn't read any of those accounts at the time. The idea that it could turn a newspaper reporter into an amazingly talented artist for a few hours was very impressive, but what might it mean for someone like me? Osborne's experience sounded like a medical parlor trick, and not exactly life changing. Could TMS actually remediate autistic disability, as Lindsay and Alvaro were suggesting? And if so, would it last?

Alvaro explained that each area of the brain does something different, and any part of the outer surface can be targeted with focused TMS. Stimulation of the visual cortex at the back of the head might make me see stars, while stimulation of areas in the front of the brain could remediate depression. He said the study they were inviting me to join would focus on five small areas in the frontal lobe, any of which could affect emotion or language. "We think those areas may work differently in people with autism," he said. When I asked for more details about the target areas, Alvaro said that I would have to wait until the study was complete.

The frontal lobe is involved in many functions, like speech, reasoning, decision making, and many of our deeper thought processes. "Stimulations just an inch apart can have very different effects," he explained. "The impact of TMS can be complex because the stimulation of one area can spill over into other interconnected regions, and the results may be hard to predict." Alvaro said that our brains are interconnected from side to side, so a stimulation that "turns up" an area on the left side may "turn down" the corresponding area on the right. Or at least that's what he and the other researchers predicted. And TMS could stimulate areas as small as 1 percent of total brain mass. "There's no other technique that can do that today," Alvaro told me.

It was time to get right to the point.

"So how do you propose to use this technology to help autistic people like me?"

His answer was simple, and it stunned me. "People with autism have trouble reading the unspoken signals of others. The conventional wisdom says they don't have the wiring in their brains to do it. Some researchers think autistic people have too many brain connections and they're all jumbled up. Others think they have the wrong connections, and that the wiring isn't there. We think the wiring is in there, but it's not working. I'm hoping we can use TMS to activate the paths for emotion and bring that sense to life. We think those networks are in your frontal lobe, and we've got several parts of it targeted for study."

If I understood him correctly, Alvaro was suggesting that I might have the ability to experience what I'd only dreamed of—that it had been there all along in my mind, waiting to be unblocked. I couldn't wait to find out for myself.

"You won't have long to wait," Lindsay told me. "We'll be starting in the next few weeks."

They had invited me to dinner in the hope that I might tell a few people about their research, which could help them find volunteers for the study. I'd agreed to talk because my curiosity was piqued. By the time we'd finished our dessert I had not only decided that theirs was a story that had to be shared, I had resolved to take part in their experiments myself.

Why Change?

"WHAT'S WRONG with how you are now?" That's what Martha and Cubby asked me as I immersed myself in reading about TMS research. "You're fine the way you are." I realized there was a large gap between the image I had of myself—an oblivious, insensitive, unwittingly arrogant social failure—and the image others had of me as a successful business owner, auto enthusiast, family man, and author.

There were even people who saw me as a photographer, a creative artist of sorts. I'd taken up image making as a hobby when my son was little, and I seemed to have a knack for it. Performers held a particular fascination for me, and in a sense, photography became a second chance for me in music and entertainment. By the time my son was a teenager, musicians, circus performers, and even some of the big state fairs used my images. Cubby even worked as my assistant at times. With all that apparent success, I can understand why people found it hard to believe that I considered myself a failure, but there it was.

I got my first glimpse of that difference in perception when my book came out and people who knew me read it. Bob Jeffway is one of my oldest friends. We met back in the late 1970s, when we designed electronic games together at Milton Bradley, and thanks to our shared interests and eccentricities we've stayed close all these years. Bob's wife, Celeste, had a particularly memorable reaction. "Jeez, John," she said. "We've known you thirty years and we had no idea you felt like you said in that book. You always seem so confident and secure. . . ." Then Bob told me that he'd also been bullied as a kid, and just the memory of that time was almost enough to make him cry. Yet you'd never have guessed that, to look at him today. I was beginning to realize that none of us knows what another's life is like.

"What if their machine changes you and you don't like us anymore?" That was Martha's fear. My second wife lived with chronic depression, and she was always able to see the downside of things. I'd gotten used to that and had often wondered whether depressed people might see things more as they are rather than how we might wish they would be. I asked myself if what she feared might come true, but I didn't see how a brain stimulation could make me fond of someone I formerly despised or cause me to turn without warning on my friends.

Like many teens, Cubby seemed mostly indifferent. At eighteen, he was wrapped up in his girlfriend and his chemistry studies and cared far more about the latest organic compounds than anything I might do or say. Martha and I were mostly a source of money, rides, and cellphone service. But there had been many times in my own life when other people had looked at me and said, "You don't care," or "You're just totally indifferent," when in fact I cared very much. One of the things I'd learned about autism was that it caused me to appear indifferent even when I had very strong feelings about something, and I had to wonder if Cubby was the same as me in that regard. Maybe he really was indifferent, or maybe I just couldn't tell.

Then there was my ex-wife, Little Bear. Cubby had told her about the research when he saw her—which told me he had been listening after all! She was still dubious about my autism diagnosis, saying it was just my excuse for bad behavior. The fact that I'd been tested didn't mean much to her, and the theories of Alvaro and his team meant even less. Whenever I talked to her about autism, she became angry or dismissive. A more emotionally intuitive person might have seen that as a clue that my words were hitting a little too close to home for her, but all I knew was that she yelled at me, so I backed away. Five years later—in a strange twist of fate—she would learn that she too was on the autism spectrum, but that insight and her acceptance of it were still some way in the future.

Cubby's mom and I had been divorced a few years at this point, and we had a contentious relationship. That was partly because I'd started dating Martha while Little Bear and I were still together and it didn't help that we'd subsequently gotten married and built a house in Amherst, where Cubby eventually chose to live full-time. We'd moved to Amherst for the schools, but Cubby ended up dropping out in eleventh grade, just as I had done thirty years before. He still stayed with Martha and me almost every night, and his mom didn't like that one little bit.

When I was Cubby's age, my all-consuming passion had been electronics. For my son, it was organic chemistry. When he was six, I was afraid he'd never learn to read. When he was ten, school psychologists talked of his major learning challenges. Now, on the cusp of adulthood, our son had surmounted many of those challenges and developed an obsession with science.

His interest was so strong that it had caused a rift in our family. Kids who love organic chemistry invariably explore one of two things—drugs or explosives. Cubby chose explosives. His mom had gotten him started with her own interest in model rockets. By the time he turned seventeen, Cubby had taken over our garage with his test tubes and chemicals.

His mom was furious, saying I had lured him away from her, but that wasn't true. In fact, Cubby and I fought so much about his mad-scientist experiments that he was talking about moving his lab to his mom's after virtually abandoning her house a few years before. We argued about that a lot, and the TMS study was a welcome distraction from our nonstop bickering.

Little Bear—I still called her that—was skeptical of most things I did, TMS included. "I don't see the point of it," she said.

The distinct lack of enthusiasm from my family made me reconsider the wisdom of my intended actions, but only for a moment. My desire for self-improvement was incredibly strong. I'd heard a lifetime of "This is how you are." Then I learned I was autistic, and the talk changed to "There is no cure for autism." I'd made the best of things because it was all I could do. I'd even come to see that I had great gifts by virtue of my differences and that my social disability was just one facet of who I was.

Martha said I had a kind of "sad acceptance" of the reality of autism in my life. Just as I lived with my autism, she lived with depression. Everything in her world was sad, so in that sense we seemed to go together. She had tried a dozen drugs to get out from under her depression, but they only lifted the veil partway. In my writing and speaking I had said that my autism was a way of being, that it was part of who I was. That it wasn't a disease and there was no need for a cure. I still believed that, but I also believed in being the best I could be, particularly by addressing the social blindness that had caused me the most pain throughout my life.

So as I prepared to undergo the treatment, Martha was fearful, Cubby appeared indifferent, and Little Bear was openly dubious about the whole endeavor. But once I'd seen the possibility of change and improvement, my drive to try it was unstoppable.

Horsepower

AS I WAITED for the actual TMS study to begin, I became obsessed with trying to piece together an understanding of how the machinery worked based on my knowledge of electronics and the things Alvaro and Lindsay had explained. I'd left our dinner with a long list of recommended reading and the invitation to email or call them with any questions. Free tech support for the brain—it was a smart offer on their part, making the whole prospect more appealing to me.

My electronics background had the benefit of making TMS feel familiar and less threatening than it might have been to someone else. Still, I tend to be an anxious person, and so I tried to distract myself from potential downsides by ruminating on the technical aspects and on how I might make a truly meaningful contribution to their work. Never mind whether the researchers actually wanted my assistance—in my autistic way I didn't give that a moment's thought.

My friend Dave Rifken, a radiologist at our local hospital, was very interested in the study. The two of us—a pillar of the community and a social outlaw—were an unlikely pair, but we'd bonded over a shared love of Land Rovers and the outdoors. We had been friends for nearly ten years, and now, for the first time, we had a medical topic to talk about, in contrast to our usual banter about Land Rovers, Jeeps, and off-road driving.

My incipient involvement in medical research was right up Dave's alley. "It's just fascinating what those guys want to do," he told me. "Did they tell you what areas they want to stimulate?"

Alvaro had said they had several possible targets in the frontal lobe, one or all of which might contribute to unlocking emotional insight in me. And there was a chance that none of them would do anything. That was the nature of research, he had explained. But the main reason for my ignorance was intentional. As Alvaro told me, "I don't want to evade your questions about the exact areas we target, but I'm afraid to tell you too much because I don't want to risk that your responses to TMS will be affected by knowledge of what we are doing. We have to do our best to ensure that your responses are the result of the TMS and not affected by the power of suggestion. The human brain has a remarkable ability to change itself, and I don't want to tell you things that might set your mind on a path that would alter the results in our study."

When we'd spoken at dinner he'd given me an example of how this could happen. One group of students in a research study was told they were exceptionally smart and gifted. Another group of students was told they were just average. Even though the two groups started out matched, the "smart" students outperformed the "average" ones.

That difference was all in their minds, but the better test scores were real. Hearing that, I agreed that Alvaro's decision to keep me in the dark about the specifics made sense, and I'd also agreed that I wouldn't ask about the exact areas that were stimulated until we were done, or reveal anything I learned about those areas to anyone else. "We're happy to explain it all to you afterward," he assured me, and I began looking forward to that moment.

When I repeated that to Dave, he said, "You'd better start studying the brain, so you can make sense of the explanation when you get it." I agreed with him, but I was also captivated by the other side of TMS—the electronic technology. Dave was dubious of my intensive interest there.

"What's a schematic diagram of the machine going to tell you about the treatment?" he asked me. "Figuring out how the stimulator works isn't going to tell you any more about what it does in your head than taking apart an X-ray machine will tell you about setting a broken leg."

With some chagrin, I realized he was right, but it didn't stop my pondering. I wanted to be useful to the project as more than a guinea pig, and the technology seemed a promising pathway.

In search of a more receptive ear, I turned to my friend Bob Jeffway. As an engineer, he had a totally opposite view, one that was considerably more adventurous. "Of course the technology is important. Let's figure out how it works. Maybe we can design a home version and try it ourselves. Who needs doctors?"

Bob was very good at imagining ways in which cool and interesting million-dollar electronic systems could be reduced to their $39.95 essentials and sold in Toys"R"Us. We'd pulled that off together when we worked at Milton Bradley, building the first toys with electronic speech, and he'd continued doing projects like that on his own ever since.

Entertaining as the prospect was, I didn't need a do-it-yourself version. Lindsay had already offered me a free ride at the lab. "Maybe later," I told Bob, and went back to exploring what the TMS machine would have to be like to deliver the kind of magnetic pulses that could actually change my brain.

A few calculations and a little bit of reading told me they needed some serious power to do what they proposed. There were two reasons for that. First, the wires between brain cells are tiny, and it's very difficult to induce a big charge in a tiny object. Electromagnetic energy prefers big coils of copper wire to microscopic threads of conductive brain tissue. Another complication would be the distance between the TMS coil and the brain cells the researchers wanted to energize. Radiated energy—whether it's light, sound, or magnetism—dissipates rapidly as you move away from the source. Shine a penlight into your eye and it's unbearably bright. Shine the same light into a big warehouse and it barely penetrates the darkness.

Double the distance equals one-fourth the illumination; that was the rule of thumb we used in my music and theater days, and I realized it was generally applicable to TMS as well. You can see this principle clearly with a light meter, a tool professional photographers and stage designers use daily. If the light meter reads 100 when you stand 10 feet from a light, it will read 25 when you move back 10 more feet. With magnetic fields, the energy also diffuses with distance, like ripples on a pond.

Although the distances in TMS would be much smaller—fractions of an inch—I realized that the principle would be the same and the effect even more dramatic. In music we measured the strength of a loudspeaker magnet at a distance of a millimeter from the surface. At that distance, its strength was quite impressive. But just 2 millimeters away, it was only a quarter as strong. Four millimeters back, the strength was down to one-sixteenth. Get back 25 millimeters—a mere inch—and the strongest loudspeaker's magnetic field was reduced to almost nothing.

This is why magnets in a high-performance sound system had to be so big and why the coils in the loudspeakers had to be so close to the magnets. A big concert sound system could have a couple hundred speakers with several thousand pounds of high-performance magnets at their core. Everything fitted together very tightly to stay in the confines of the con-centrated magnetic fields.

TMS would be a lot harder to execute successfully, because the magnetic field had to reach through my hair, skin, and skull. That was at least a half inch—way more distance than any loudspeaker system. If you put a speaker coil that far from its magnet it wouldn't work at all. That made me realize that the magnets for TMS had to be orders of magnitude more powerful than anything I had used in music.

We'd packed a lot of punch into our amplifiers, but you'd have to ramp that up a hundredfold to deliver electromagnetic energy across a half-inch gap. That was some serious power—truly lightning in a bottle—which made me wonder how safe the whole proposition was. The TMS machine was certainly going to be generating thousands of volts to get the kind of power they would need. But I told myself that electrical safety is of paramount importance when designing gear for a hospital. I had to trust that the TMS machine designers had done their job and that all of us would be safe. I resolved to treat the equipment with a healthy dose of respect.

My mind wandered as I tried to imagine what the actual sessions would be like. The power levels would be high enough that the coil would make an audible pop when it fired. Might that sound affect my response? What about heat? The magnetic fields would warm my brain cells to some degree. Microwave ovens use a similar principle, only in concentrated form, but they probably didn't plan to cook me in the TMS lab. There were all sorts of ways I could have talked myself out of going through with the study, but instead I jumped in headfirst, with hardly a moment's hesitation. Alvaro was a professor at Harvard and the head of a brain center at one of the top hospitals in the country. Both he and Lindsay were obviously bright and dedicated. If anyone had the resources to get and use safe, top-notch lab equipment it would surely be them. And as Alvaro had told me at dinner, he'd been doing TMS for almost twenty years.

And most of all, the goal they were pursuing was very dear to me. I resolved to believe and trust them. They would not fry me or my brain. We would go down this road together—they as the experimenters and me as the lab rat—and we would see what that energy did when it got inside my head.

Informed Consent

WHEN WE WERE at the restaurant, Alvaro and Lindsay had invited to me to visit their lab before the study began, and of course I took them up on it. I'd looked at pictures of TMS equipment online and studied how it worked. The next step was to observe the scientists in their den. Later in the week following our dinner I made the drive from Amherst to Boston. When I exited the turnpike, emerging in a valley of strip malls and car lots, I was afraid I'd taken a wrong turn. Eventually parkland opened up on both sides of the street and I saw tall buildings towering ahead of me. I drove through a very impressive hospital campus before finding the entrance for Beth Israel and turning in.

A receptionist gave me directions to the brain center, and after walking for what seemed like half a mile I finally reached my destination—the Berenson-Allen Center for Noninvasive Brain Stimulation. The walls of the quiet waiting area were lined with framed magazine articles describing the work the scientists had done, the acclaim they had earned, and the patents they'd been awarded.

No one had noticed my entrance, so I walked down the hall and gazed into the open doorways, hoping to learn something.

"There you are!" I turned to see Lindsay approaching rapidly, with two people tailing her. I resisted the sudden urge to run. "This is Shirley Fecteau," she said, pointing to her dark-haired colleague. "She's another postdoc. And this is Lin, our lab assistant." I photographed their name tags with my iPhone because I have trouble with names and that seemed like the best way to keep track of who was who. They looked at me a bit strangely and herded me toward a room at the far end of the hall. We all took seats, and they looked at me expectantly.

"Alvaro is on his way," Lindsay said. I thought I had snuck into the office quietly, but her words made me realize that the receptionist must have called ahead.

Never before had I been accorded such respect and deference in a medical office, treatment I imagined was usually reserved for famous surgeons visiting from afar. I was wondering what I'd done to merit such an honor, and then Lin mentioned my book. Not only was I the first autistic adult to join their study, I was the first autistic person any of them had met who had written a book about the experience.

The researchers were young, and most of their work to date had been with autistic kids and young adults. They hadn't met many successful older autistic adults, and in the spring of 2008, my book was pretty novel. Consequently, I wondered if they were waiting for me to say or do something. I thought of the sixties soul singer Archie Bell up onstage, shouting, "We don't only sing, but we dance just as good as we want!"* Unfortunately, I didn't sing, and I sure couldn't dance.

Years before, my grandmother had taught me to ask people about themselves when I didn't have anything else to say, and it generally worked. I asked Shirley, who spoke with a strong accent, where she was from. It turned out she was French Canadian. I already knew Lindsay had studied at UCSD, and I learned that Shirley had come to Alvaro's lab from Quebec. Scientists seemed to travel great distances to work with him, and I asked Lindsay and Shirley why that might be.

---

* This passage comes from the hit song "Tighten Up," written by Archie Bell and Billy Buttier. The song was a big hit in 1968 for Archie Bell and the Drells of Houston. Several of the bands I toured with played the song in the seventies.

It turned out that there were not very many people doing research into noninvasive brain stimulation, and Alvaro was a leader in the field. In the years to come I would meet visitors from many other countries in the halls of the TMS lab.

All of them had unique areas of interest. Shirley's interest was brain stimulation, and it just happened that the current study was on autism. But she was also interested in using TMS to treat addiction and in working on what she called "military applications of brain stimulation"—whatever that might mean. Lindsay, on the other hand, had come to the lab specifically to study autism and TMS together. Lin, the assistant, was a student at a local university and was gathering experience before deciding what to do in grad school.

Just then, Alvaro appeared in the doorway. "Welcome to the center," he said, smiling and holding out his hand. "Please join me, and let's have a look around." Walking a short distance down the hall, he motioned us into a room that was larger than the office we'd just been in, with a big comfortable-looking chair and a bunch of medical electronics.

"This is a TMS machine," he said as he patted a large box with MAG-STIM printed in block letters on its side. Then he picked up a heavy insulated cord with a plastic figure eight at the end about the size of my hand. "This is the coil," he said.

He proceeded to show me other equipment in the lab, which included brain-wave-monitoring gear, otherwise known as an EEG system, another smaller TMS machine, several more oddly shaped coils—"They have different stimulation patterns," he explained—a monitoring camera, and two computers. The only sign that this was a medical office was the soap dispenser and medical wastebasket by the sink.

It took a moment for me to take it all in, but then I started asking questions. They told me the TMS equipment was used in many studies, and it was also being used to treat patients in the hospital for depression and stroke and for other experimental procedures. Alvaro's center was exploring autism, Fragile X, depression, Alzheimer's, and a host of other maladies. The place was really hopping.

"Alvaro is not just a researcher. He's also a practicing neurologist,"

Lindsay said with some pride. "He does rounds in the hospital and sees patients here in the clinic. He has a particular focus on TMS, but he treats people with all sorts of neurological problems." There were several other neurologists in his clinic too, along with research scientists. The lab was even more impressive than what I'd envisioned—spotless and efficient and bustling with motivated people.

It was all a bit intimidating, but at the same time I felt privileged to be part of their efforts, even if I was only a glorified guinea pig. But unlike a guinea pig, I did have to willingly agree to be involved and sign my life away, after a fashion. I also had to be tested—to make sure I was "fit for study." All that happened on my next visit, two weeks later.

Lindsay met me in the Beth Israel lobby and led me into an examination room in Alvaro's Berenson-Allen Center. She took out a big folder containing what looked like an inch-thick stack of papers. "This is what's called an informed-consent form," she told me as I picked up the first of several multipage documents. It's my nature to actually read that kind of paperwork and even ask questions. Later, Lindsay would say I asked more questions than any research subject she'd ever encountered, but she said it with a smile so I figured it was okay.

After a quick read through the stack of papers I got the gist of what they said:

> We (the folks at the hospital) are going to do certain experiments on you.
>
> They may make you better, or they may not.
>
> You understand the possible risks and agree that it's okay to proceed.

It was a little ominous, when I thought about it. Maybe that was why they had taken five pages to say what they could have said in three sentences—to ensure their inscrutability to the average person. But mostly I felt a little disappointed. The scientists had talked about "remediating the deficits of autism." That had had me imagining life-altering changes, but there was nothing at all promised in those pages.

"We're going to do at least six stimulations," Lindsay explained, "and we'll have you do tests on the computer before and after the TMS so we can measure the effect."

"At least six?" That sounded a little open-ended, but she was quick to clarify what she meant. "There's a possibility of error doing these stimulations. If something doesn't work out, we might have to redo it." I wondered what she'd meant by "error," but I guessed it made sense. This was, after all, research.

"Has Alvaro ever done something like this before?" I asked.

"This is a brand-new study," she explained, "but he's based it on his previous TMS work in other areas of the brain, and on animal models." Hearing that I might be Subject One was a little scary, but it was also exciting.

When I arrived at the TMS lab in March 2008, I was told there were already five other volunteers for the study. Two had beaten me to the hospital and signed their consent papers, but no one had actually started the stimulations. In the next few weeks three more people would join this first Harvard–Beth Israel TMS autism study, which would run through the summer. None of us knew one another, with one exception.

Michael Wilcox was a little older than I was, and very smart. He was a former financial analyst who had gotten fed up with corporate life and retired to a farm in the Berkshires. Then, at about age sixty, he learned he was autistic. We'd met when he read about my first book and realized I was local. He emailed me to ask where he could find it, but it wasn't yet on sale. "You may be able to buy a review copy on eBay," I'd replied, and that's what he did. After reading the book, he invited me to lunch, beginning a tradition of periodically shared food and conversation that continues to this day. We both joined our local Asperger's support group and took an immediate liking to each other. A few weeks earlier—after the dinner with Lindsay and Alvaro—I had told the group I'd decided to try TMS, and he said, "I think I'll talk to them too."

A week later he told me he'd done just that and shared how impressed he was with Lindsay and Alvaro and their ideas. *He's just as excited about this as I am,* I realized. Michael and I had agreed to do our sessions at different times, and we resolved to avoid talking to each other as it was happening, because we didn't want to affect the results.

It didn't completely turn out that way, but we did our best.

The description of the research in the consent papers threw some cold water on my eagerness. According to the forms, they were going to test our ability to recognize common objects. And they were going to measure prosody—the rhythm, stress, and range of my voice as I expressed emotion—to determine whether TMS changed those things.

It wasn't the most thrilling proposal, but I told myself they had to take small steps before the big leaps. I did wonder if they thought TMS would do a lot more than the forms suggested, and later on I asked Lindsay if my suspicion had been true, and what her real goal was. "I want to win the Nobel Prize for my discoveries in autism!" Her enthusiasm made me smile, and I hope that comes true.

The potential risks outlined in the waiver didn't seem too scary, and the forms explained that "previous studies have shown that effects of TMS lasted half as long as the time of stimulation." That meant a half-hour stimulation would only affect me for fifteen minutes.

Although the briefness of the experiment's effect seemed somehow reassuring, it also worried me. Where was the lasting benefit? At my next opportunity, I asked Alvaro what good TMS could do if its impact was so short-lived. "We believe its effects are cumulative," he explained. "We think TMS opens up paths in the mind, and when you use them, you make them wider. We also believe continued stimulation strengthens them. Think of sledding down a hill when you were a kid. When the snow was fresh, you could make a path anywhere. But after a while the paths get worn in, and all you can do is follow one or two routes down the hill. We think TMS can open new paths and help tread them down so you can keep them open. Some of the people in our depression program are seeing benefits lasting six weeks to two months." Their TMS depression program was very close to getting FDA approval, and they were excited by that prospect.

"We've done TMS for depression for a few years in Spain, and they do it in Canada and elsewhere in Europe. It's not permanent, but it lasts a long time. It's a good alternative to daily medication for the people it helps."

"How similar are the depression and autism therapies?" I asked. Penicillin will treat an infection in your ear or your toe and all you have to do is swallow a pill. I wondered if TMS worked the same way—going wherever it was needed.

"They're not similar at all," Lindsay told me. "Remember, TMS only reaches a tiny part of the brain. It's not like a medicine that diffuses throughout the body. The areas that are stimulated to treat depression are totally different from the parts we're stimulating in the autism study."

Hearing her made me realize that my penicillin analogy was wrong because it referred to a pill. TMS might better be compared to an antibiotic ointment—you could put it on your cheek or on your leg, and it would treat infection in either place. But putting it on one spot wouldn't do anything for an infection somewhere else.

"That's a better way to look at it," Lindsay agreed when I suggested the comparison. I'm very lucky she was patient as I tried to turn complex questions of neuroscience into simple analogies a bumpkin like me could understand. Then I wondered if the study we were about to begin would lead to an FDA-approved treatment. "We've got a long way to go for that," she told me. "First we'd have to demonstrate a real benefit in this study. Then we'd have to do several follow-up studies to refine a treatment. Then we'd need to do a bigger trial with lots more people. This is just a small study with a few subjects. Getting from the lab to an FDA-approved treatment is a very slow process."

As it turned out, TMS would be approved by the FDA for depression while we were doing the TMS autism study. There was considerable rejoicing in the lab when that happened and they realized their research had finally paid off. Today, Alvaro's lab has an outpatient clinic that's using TMS to treat a steady stream of patients. His reputation is so good that patients travel hundreds of miles for their treatments.

Their talk of success with depression was reassuring. Even so, other people's reactions sometimes got me worried. One day I told someone about TMS, and he said it sounded like another term for ECT, the electro-shock therapy of medical horror movies.

"I don't think that's true," I had replied hesitantly, but I didn't know enough to really explain the difference. Later, Lindsay helped me understand. "ECT works by putting so much energy into the brain that it induces a seizure, a massive reset if you will. ECT is also very diffuse, whereas TMS is tightly focused. Even today, ECT is a violent process, enough so that it's done under anesthesia. TMS is none of those things, and its effects are a lot gentler. The two technologies both deliver energy to the brain, but the energy TMS puts in the brain is a tiny fraction of what's used in ECT and it's aimed right where we want it, not scattered everywhere."

When I read about ECT—and looked at the energy levels involved— I realized that ECT was probably "burning out" some of the delicate wiring in the brain, which was scary to contemplate. From the beginning, ECT had a reputation for wiping away pieces of patients' minds. TMS and ECT were both electrical therapies, but one was at the level of an AA battery in your pocket flashlight, while the other was like the high-tension wires coming out of the generator station at Hoover Dam.

That really put the two techniques in perspective, and a little later Alvaro reinforced what Lindsay had said. "We've done everything we can to ensure this is safe. There's always a neurologist on duty when we do TMS, and no one has ever had a seizure from it in this lab. The energy levels we use for TMS today are minuscule compared to what doctors used for ECT and a fraction of what they used for direct stimulation treatments twenty years ago. We are doing depressive stimulation, which is inherently safer than excitatory TMS. It's always safer to turn something down than to turn it up."

The difference, as they explained it, was that depressive stimulation meant producing a slowing down or weakening of brain activity in a particular area, whereas excitatory TMS meant speeding up activity in the target part of the brain. But if all they performed on us were depressive stimulations, how could they actually improve anything? Musicians turn it up to make rock and roll better and race car drivers go ever faster. Turning things down seemed like a road to nowhere, but Alvaro had the answer.

"The brain has a huge network of wiring called the corpus callosum that keeps the two sides of your brain connected and in balance. One way to think about that in the context of TMS would be to imagine a teeter-totter. We can raise a side by lifting it up. But we can also raise that side by pushing the other side down. We think TMS works the same way, and it's why we can use depressive TMS to lift or lower. By pushing one side down, we think the corresponding area on the opposite side will rise. That's safer than lifting directly, and that's what we are doing here." And as an additional salve for my concerns, Lindsay added, "The institutional review board goes over every detail of proposed research, and they are very cautious about what they approve."

The review board was part of an eerily titled organization called the Human Subjects Protection Office. That name alone would make anyone uneasy, but I chose to trust the scientists. I knew of course that there were risks; with this sort of research there's always a leap of faith required, from both the researchers and the subjects. That's how science moves forward. And who better to do the leaping than those who feel they might benefit most from the results—those whom researchers call the "affected population"?

Looking back, I was about the furthest from "informed" that one could be, in spite of what the doctors had told me and all my outside reading. None of us knew where the results might lead, and there was no way to know if TMS would even make me better. To complicate things further, I realized I couldn't have even defined what "better" meant, had someone at the hospital asked me. It was almost a shot in the dark. But if it did help, I was getting in on the ground floor and I would be a decade ahead of everyone else who might eventually benefit from the treatment. I took a deep breath and signed on the dotted line.

The History of
Brain Stimulation

I MAY HAVE BEEN AWED into submission by my visit, but I still retained the smallest trace of skepticism, even as I was deciding to trust my brains to this team of smart researchers who might be planning something really great, or possibly something a lot worse than a painful shot.

I also retained a measure of what some would call arrogant confidence when it came to the electronics. After all, with all the work I'd done with transformers, inductors, and electromagnetic pulses, I figured I might well know more about how TMS delivered energy than they did. That was the one area where they didn't have all the answers when I visited the lab. I had asked if their devices used tubes or thyristors, and no one knew. I asked what the peak voltages were, and they weren't sure of that either, though they knew they were high. By that time, they had wowed me with so much that they *did* know, and with their level of intellect, that I was desperate to find even one area where my own knowledge might be enough to stand alongside theirs.

If there was such a place, it would be the electrical engineering side of TMS. I readily conceded that I had absolutely no idea where to deliver a stimulation, or in what quantity. All I knew were the general parameters.

Too little energy and nothing would happen. Too much and those tiny brain wires would go up in a flash. *Watch the ears when you try it,* I told myself. Little white tendrils emanating from the holes would be a warning, for sure. Smoke was always a reliable indicator of electronic circuit overload.

I settled in to consider what they were doing. I had to take my mind back to the 1980s, when I worked on power systems that delivered the kinds of pulses Alvaro's people wanted to use for TMS. The pulses we created back then were used to fire laser tubes, or "other things" in classified weapons systems. Alvaro and Lindsay had talked about high-power pulses in their medical machines, but I'd worked with high-power pulses too. When I worked at Candela we powered lasers that shot light to the moon and vaporized crucibles of uranium in microseconds. When I'd worked at Isoreg I'd designed power conditioners to withstand nuclear blasts, and you can't get much more high-powered than that. I had a good understanding of how electromagnetic energy worked, but twenty years had passed. Now I struggled to bring it all back.

Electricity and electromagnetism are two things you can never see. You just have to imagine them and observe their effects. We can bind two objects with a piece of rope, and it's obvious what holds them together. Make one of those objects a piece of iron and the other a magnet and they will snap together tightly even though there's nothing visibly holding them together. If the magnet is strong enough, you'll break your back trying to pull them apart.

Tie a rope between two cars and you can pull one of them out of a mudhole or a snowbank. Use a wire "rope" to tie a generator to the power panel on a big building, and you can deliver light to a thousand people. Which carries more energy—the rope or the wire? Clearly, the wire has the edge.

I've always felt an affinity for powerful elemental forces. Electromagnetic induction is one of the things that makes modern society possible. It's the principle by which electricity is generated in power plants all over the world, and it's the principle behind electric motors in toys, trains, blenders, and washing machines. It makes Tasers tase and electric guitars sing.

The initial theory of electromagnetism was arguably the greatest achievement of Michael Faraday, a prolific self-taught British scientist of the early nineteenth century. In the fall of 1831 Faraday took a ring of soft iron and wrapped copper wire around it. He called that helix A. Then he wrapped another length of wire around the same ring and called that helix B. He wired helix B to a meter and hooked the wires from helix A to a battery.

You have to ask yourself what possessed Faraday to do such a thing when it had never been done before. What did he think would happen? Sometimes the most remarkable inventions spring from inexplicable flights of fancy.

Faraday would have said, "We have to start somewhere," and that's what he did. As soon as he touched the wires from helix A to the battery, the meter needle jumped. When he took the wires off, it jumped again, this time in the opposite direction. That was the first demonstration of electromagnetic induction—the process of turning electrical energy into magnetic energy. To an engineer, Faraday's achievement was akin to inventing the paper clip, or maybe the wheel. It was an incredibly big deal.

And it hasn't changed much in 160 years. If you were to hold Faraday's helix up to a TMS coil in Alvaro's lab, the similarity would be instantly apparent. And best of all, it still works the same way. That's what I like about engineering. When you figure out a principle, you can always count on it. When I repeated Faraday's experiments in my basement lab as a teenager, I felt my own thrill of discovery.

Faraday also learned that he could wrap wire around an iron core and induce electricity in a freestanding coil next to the wire-wrapped core. That proved that energy moved through space—it didn't just travel through the iron bar. That property is what makes TMS possible today.

In TMS there is one coil of wire wrapped around the iron core, which is held against your head. The "receiving" coil—if you could call it that—is the invisible jumble of biological "wires" that connect the neurons of the brain beneath where the TMS coil is placed against the scalp. Scientists put pulses of electricity into the TMS coil, creating magnetic fields that induce electricity in the wires of the brain.

So the question we come to now is: Where does it go from there?

The principles of electromagnetism told me that the physical orientation of the wires in my brain would affect their sensitivity to TMS. Brain wires that go up and down might not receive a signal that would be strong on side-to-side wires, and vice versa. So which way did the wires in my head go? You can look at a computer circuit board and see which way its wires go. Brains aren't so easily examined, and their wiring is immeasurably more complex. Every person is unique, but I figured there must be general patterns of brain wiring, just as our bodies are generally similar in other ways. I turned to the literature for an answer, but I could not find a consensus.

Some scientists—including Alvaro—believe that brain wiring runs in all directions, and they counted on that randomness to ensure that some energy would be delivered beneath the coil, no matter how it was placed on my head. Another researcher—Dr. Manny Casanova at the University of Louisville—had published an article suggesting that the cortex was dominated by structures called minicolumns, stacks of neurons that make up the brain's outer layer. He seemed to believe that the minicolumns have a core of excitatory neurons surrounded by a sheath of inhibitory "horsetail" neurons. The inhibitory neurons—by virtue of their structure—are more sensitive to the effects of a TMS coil laid on the scalp, when the pulses come slowly. Casanova suggests that greater sensitivity makes most TMS inhibitory in nature. He believes higher-frequency stimulation can reach through that inhibitory curtain to stimulate the excitatory neurons at the core of the minicolumns, and that explains the working of a different type of TMS. However, other scientists disagree with that notion; a positive answer will have to wait for advances in brain imaging technology.

One thing the researchers all agreed on was that some of the neuron wires were short while others were quite long—reaching right across the brain and even out into the body's nervous system. So TMS might deliver energy to the front of my brain, but those long wires could carry it everywhere else, and who knew what that would mean. It was a lot to wrap my head around, and it showed how incredibly complex the brain is and how little we really know of its architecture. Even today, the threads that connect our billions of neurons together are only mapped in the broadest, most general sense. Details of the individual connections remain largely unknown.

The research papers I tried to read were full of technical terms that I had to look up and still didn't always understand. The problem was that I didn't know enough about the architecture of the brain to make much sense of them. Dr. Casanova had described wiring between the minicolumn structures as if it were clear as day, but when I looked at images of actual brain tissue from his article, the threads seemed to go every which way, and it was hard to call what I saw anything but random.

Then there was the question of autism. For years I had heard people say things like "Autistic brains are wired differently," and I wondered how literally true that might be. Do the wires actually follow different pathways in brains like mine, and are there individual differences in up-and-down or side-to-side wiring? I learned that the limited research we've done with brain tissue hasn't shown huge differences, but that might not be the whole story. There was a lot of talk about connections, and there seemed to be several schools of thought. Some researchers thought autistic people had too many connections, while others thought we had too few. One team of researchers—Nancy Minshew of the University of Pittsburgh and Marcel Adam Just of Carnegie Mellon University—proposed that we had both in a complex but fascinating theory. Just and Minshew had used brain scanners to map out some of the major connection pathways in the brain. They hypothesized that autistics have imbalances between the pathways, but interestingly, they thought the balance might change as we age. Our "different connectivity" might make for mental confusion, but it might also help some of us excel at reasoning tasks that use one concentrated area of the brain. Their ideas were extremely interesting, but I found it hard to make the leaps of reasoning to connect microscopic differences in brain structure with real, observable behaviors.

Dr. Casanova thought the minicolumn structures that make up the cortex are different in autistic people too, but his explanation of what that might mean was too technical for me to grasp. After overloading my brain with online reading, I wrote to him and asked a few questions, and to my surprise he responded right away.

Manny has been studying autism for about ten years. He started out in pathology, looking at brain tissue samples from people who had died. What he learned there led him into the lab and toward novel lines of research with living people. He's been using TMS most of that time. He had his theories, but he didn't know how it works with any more certainty than Alvaro, and the two of them don't always seem to agree. What they did agree on was this: medical imaging—as good as it is—is not at the point where the effects of TMS can be observed at a cellular level in living people, and until that changes, we are left with observations of larger brain function and best hypotheses.

I asked Dave what he thought about my struggles to make sense of what I was reading, and he laughed and said, "I'm just a country radiologist! You probably know more about it than me, with all you've been reading." Maybe so, but I was a long way from figuring it out.

The differences between the scientists' theories just served to illustrate that while everyone agreed on how TMS energy is made, there are differing ideas of what happens when it gets into the brain. Did the TMS pulses overwhelm the signals that flowed on the interneuron circuits? Did TMS energy cause the neurons to shut down or become paralyzed by electrical overload? Or might it make them hyperactive? All those theories and more have been proposed, and I realized that there might well be more than one explanation. The brain has many different kinds of neurons and they might respond very differently to TMS signals. If my brief foray into brain science taught me one thing, it was how little we really understand about the brain's inner processes.

That made the ideal location for the TMS coil a little uncertain, as I would soon discover. Alvaro and his team had proposed to change the way I think, but reason and cognition are two of the most amorphous and elusive things that go on inside the brain. Even with the latest imaging tools, we struggle to understand where thought resides and how it is processed. Alvaro had already told me they had several target areas in the current study. The reason they had that many was that it wasn't clear which—if any—was the exact area they were looking for. It's easy to determine where the nerve fibers from your eyes and ears enter the brain, and we know where the nerves for our hands and legs emerge. But where do we form a thought like "I love my puppy"? We don't know for sure, but there are probably many areas involved in synthesizing that four-word emotion.

Magnetic resonance imaging (MRI) has been able to show us the inside of the brain for quite a while. The newest functional imaging tools can show us activity in the brain as we think and do tasks. But its resolution is limited. Faced with a brain that holds hundreds of millions of neurons in every cubic inch, even those state-of-the-art tools can't tell us what's going on except in the most superficial way.

We know how the brain is connected to our muscles and how that system works. But where did I get the idea to move my arm just now? How do we decide to talk, or blink an eye? Those are very different matters. We have very little knowledge of how abstract thoughts are formed and how and when they are translated into action. Yet abstract thought is one of the things that makes us human.

And I couldn't help but take my line of inquiry a step further. If the brain contains the mind, does the mind contain the soul? Is there somewhere physical we can point to and say, "There is the essence of John"? If Alvaro and his team knew the answer to that, they sure didn't share it with me.

That's what makes TMS and similar experimentation such a leap of faith. We cannot truly understand the impact of any given stimulation until we try it. If things go well, all is great. But what if they don't? Would we be able to reverse a bad effect? We had dreams and theory, but we didn't have hard answers.

I had started reading about brain wiring in the hope of finding an understandable foundation for the experiments I'd volunteered to take part in. But all I got was more confusion, and I was left with the knowledge that I just had to trust Alvaro's experience and instinct.

All we had were a TMS machine, some ideas, a few volunteers, and a lot of hope. For some reason, that still felt like enough.

Mapping My Brain

A WEEK AFTER I SIGNED the consent forms, it was time to begin. "We have to start with some basic tests and an MRI of your brain," Alvaro had explained. "That will allow us to make sure you are okay for the study and target the TMS precisely where we want." What he didn't say but what I imagined he meant was "We'll also make sure there are no tumors or extra things growing in there that make you the way you are."

With some trepidation, I drove the now familiar route to Beth Israel on the afternoon of my MRI appointment. I met Shirley and Lindsay at the TMS lab first for the testing. They sat me down at a computer.

"You're going to see a person on the monitor who will say something to you," Shirley said. "All you have to do is push the front button if what they say makes sense and the back button if what they say is nonsense. Let's try a practice run."

"The sky is green," the face on the monitor said. I felt them watching as I pondered what to do. Where I came from, the sky was not generally green. I pressed the rear button for nonsense and the next question arrived.

"Eat the cake." Makes sense.

"Drink the highway." Nonsense.

"Okay," Shirley said. "That's great." But then the questions turned strange. After a few minutes, I began to wonder . . . maybe they did drink highways in Shirley's world. I thought back to my rock and roll days, and suddenly all of the questions seemed a lot more ambiguous.

"Choke me." *Okay,* I thought, *I'll say this one makes sense; I'll choke you if you want.* I pushed the front button.

"The road is bumpy." Another front button.

"The dog is blue." That made me think of the pampered dogs I'd seen in my time. I was a fan of the movie *Best in Show,* but I could not recall any blue canines on parade. Either way, it was a good movie. When I recommended it to Shirley, she told me to stay focused on the test. The words kept pouring by, some strange and others not.

"Throw me out the window." I paused for a second. This was the kind of phrase we used to program into Milton Bradley games when we set up talking demos for management. Our other favorites were "Smash me!" and "You're a real dummy!" With a small smile I remembered how our best efforts had never escaped the scrutiny of the engineering manager, and then I turned my attention to the task at hand and pressed the front button.

Finally the questions came to an end. I was sure there was some deeper purpose, and I hoped I had passed. If this was an intelligence test, it was surely a strange one. I very much wanted the TMS to work, and I was afraid my answers might disqualify me.

No one said anything, so I asked how I'd done. "Fine," Shirley told me, as if the questions and responses had been the most casual and natural conversation in the world.

It would be six months before I learned the true purpose of those strange questions. The nonsense questions were a red herring. They were actually trying to learn how much I mirrored what I heard with my body. They theorized that a prompt like "Pet the cat" would involve reaching out, so I'd move my arm quickly to the front button when I heard that. "Brush your hair," in comparison, involved a pulling back and up, so they were watching to see if it took me a tiny bit longer to overpower that pull-back impulse before extending my arm to the front button.

Mirroring had sounded like a simple concept when it was presented as seeing a smile and smiling in response. Once we got into tests like this, though, I realized how incredibly subtle and complex it was. And as for the test . . . I never did learn how I scored; I was just glad I didn't flunk out before I'd even had a chance to begin.

None of them had any idea how vulnerable I had started to feel, now that they had raised the possibility of changing my brain. I realized that my acceptance of how I was—up till now—had largely been founded on the idea that I had no other choice. You might have compared me to a prisoner who made the best of his surroundings, only to run for the gate the moment it was left ajar.

The next stop was a room where they did the brain stimulation. Shirley pointed to a chair and indicated I should sit. Right next to it I saw a large white box on a cart—the TMS machine. On my best behavior, I resisted the temptation to turn the knobs on the machine, pick up accessories on the cart, or give the thing a test run on my own. Instead, I meekly sat down and waited for what came next.

"Today we're going to measure your response to TMS. Everyone is a little different. What we're going to do is stimulate your motor cortex with a single pulse at a very low level. We use the motor cortex for this test because it's the easiest part of your brain to measure. We stimulate there, and your muscles move. We're going to find the spot that moves your index finger, decrease the TMS level until your finger doesn't twitch anymore, and then record the setting where that happens. That will help us calibrate for the actual experiments. This isn't going to have any effect on how you think or feel, because today's stimulation isn't hitting those parts of your brain. It's like an alignment session for our equipment. By measuring your brain's sensitivity in a simple test, we get an idea of how much power we'll need to use for the actual study."

"Why the index finger?" I asked her. That question elicited a very interesting response. It turns out that some of the neurons in the brain's motor cortex have microscopic tendrils that reach all the way to the index finger. Most of our appendages are connected through several nerve cells in series, like relay circuits in the spinal cord. They had chosen the index finger circuit for the directness of its wiring.

Take a moment to consider how tiny neurons are, and then imagine some of them with tiny fibers stretched three feet long. . . . That was hard to wrap my mind around, but Lindsay assured me it was true. "There are actually long stringy neurons reaching all over your body," she told me. "They are so thin that they are little more than strings of molecules, carrying the chemical and electrical signals to run our bodies."

Her words made my mind flash back to my teenage years, when Little Bear and I were growing up together. We had shared a lot of geeky interests: I introduced her to electronics, and she turned me on to science fiction. I suddenly remembered a book she gave me—*Ringworld* by Larry Niven. In it, he described something called a Sinclair Molecule Chain, a fictional wire made up of incredibly strong strands of molecules. I fantasized that these brain wires were about that thin, and I marveled at the parallel between science fiction and medical fact.

The *Ringworld* analogy somehow comforted me, and Lindsay's explanation didn't sound very scary. But then a technician, whom I hadn't noticed before, crept up beside me with wires and alcohol swabs. I knew what that meant! Alcohol and swabs were what doctors used before they broke out the scissors and scalpels, and that was not part of my plan. I must have looked alarmed because she continued: "Don't worry. Erica is just cleaning your skin. Then she's going to tape three electrodes to your hand, and we'll use a monitor to pick up the signals from your muscle activations." I relaxed again and let her stick the electrodes to my paw. It didn't hurt, but I was still temped to growl softly and bare my teeth when she leaned close. I refrained.

Truly, I was a model of good behavior.

Lindsay held the coil up to my head, and it felt cool on my scalp.

"How do you know where to put it?" I felt a little like a condemned criminal who keeps asking questions to stall his execution, but I still wanted to know.

"The motor cortex is in the same general area on everyone," she answered. Then she traced a circle on my head with her finger. "It's right here. I know about where to put the coil just by looking at you, and we'll refine the location by testing. You'll see. Okay, are you ready to go?"

Nodding my assent, I prepared to be zapped. *Pop!* I heard the machine and experienced the jolt of energy. If you've ever stuck your finger in a wall outlet, you know exactly how it felt—with one important difference. Sticking your finger in the outlet provides a zap that doesn't end till you yank your finger out. The zap of a TMS machine only lasts a few thousandths of a second. Still, that pop of electrical energy is unmistakable. You feel the energy level rise, and you feel it fall. It's a sensation akin to what it might feel like to pluck a harp string inside your head. Lindsay said she'd heard it described as a woodpecker tapping your scalp, and I guess I agree. Then she flicked a switch and the TMS machine fired out two pulses in a rapid *tap-tap.* "A doublet," I exclaimed, and Lindsay was startled. "You counted those pulses?"

"Yes," I said, "there were two."

Lindsay told me that most of the people she had studied could not distinguish the two quick pulses from one. She tried it again, with more pulses, and I still counted correctly. "That's pretty surprising," she said. "Those pulses were only separated by three-thousandths of a second." Her reaction made me wonder if I had an unusual ability to distinguish pulses, or if I was just the first person to speak up about it. Counting the pulses had seemed perfectly natural to me, just as if I were counting the beats of a snare-drum roll at a musical performance. Then I thought, *Only freaks count drumrolls,* and I decided to keep quiet.

Lindsay had no idea what to make of my ability to count fast pulses and didn't know if it mattered. I was just paying close attention to everything, because I was in a strange new environment with no sense of what might be important.

Later we would learn that my ability was a feature of autism. Some autistic people—like me—can count the beats in quick bursts of sound; while most people see fluorescent ceiling light as a continuous glow, in fact the light is flashing off and on 120 times a second. Many autistic people can see that, which explains why that type of lighting can be unsettling to some of us.

Then there was the feeling of TMS. I was surprised to feel anything at all, because one thing they had offered by way of reassurance was that the brain doesn't have pain receptors. And sure enough, what I felt in the chair wasn't what I'd call pain—it was more of an awareness of energy being fired into my head. "There are plenty of nerves in your scalp and skull," Alvaro explained, and I'm sure I felt the TMS pulse through them. But I also felt something deeper—a stirring as energy flowed into my brain.

"I don't know what to say about that," Lindsay told me. "I don't think most people feel that inner touch. I've had TMS done to me and I couldn't feel anything inside."

But the feeling inside was nothing compared to what happened next. She turned the power up on the machine and fired it again. The pulse made my arm jump like a kangaroo, but she just nodded calmly and backed the power level down. The next pulse moved my middle fingers, as if they were making me grab by remote control. The third—according to Lindsay—was right on target. My index finger gave a single subtle twitch as the TMS coil fired. They had found the spot. "Seventy-two," Shirley read the number to the lab tech, as she adjusted the power level. The next pulse was at sixty-five; then sixty.

"Frankenstein," I said, and she nodded and tried to look serious.

And that was the strangest thing. I could hear each pulse, and I could feel a twitch in my head. But at 25 percent power, nothing at all showed in the fingertip electrodes, even though I felt the zap go into my head. At 40 percent, there was still nothing. We went up slowly: 60 percent, 70 percent, then 72 percent. Suddenly a blip appeared from my fingers: 75 percent, and my fingers popped like frog legs off a dinner plate.

It was very curious. During a break, I called Dave, who said, "Great! Now you can remagnetize the strip on my credit card using your mind." It didn't occur to me that he was being sarcastic.

Next we headed over to the MRI center. I'd called Dave earlier that day to ease my anxieties about what to expect, but he'd neglected to mention that there would be a contraption to hold my head steady that resembled a football helmet with weights to hold it on the table and pads inside to keep my head from moving. He'd warned me that some people get claustrophobic, but I resolved to lie still and quiet as they slid me into the machine. "It may be easier if you close your eyes," the technician said, and I did.

"Are you okay in there?" The technician operating the machine had given me a button to push if things went wrong. I wiggled my feet to reassure myself that I wasn't strapped down and could get loose if I needed to. I'd spent time hiding in drainpipes as a kid, but this place was warm, dry, and free of snakes and critters. I was sure I could do it.

"Here we go." I heard the technicians leave the room, and a moment later the machine started humming. It played a very unusual melody of screeches, buzzes, and clicks. Judging from the sound, large masses were moving all around me, but when I opened my eyes all I could see was smooth white plastic.

"Almost done," the voice said, and then I heard the door open. Seconds later, they were sliding me out of the machine.

They'd shot 168 images of my brain, which I carried home on a computer disc that the technician handed me. "The doctor will look at the data and talk to you in a few days," he told me. The prospect of more waiting wasn't too thrilling, but there wasn't anything I could do except look at the images myself. So that's what I did.

The brain scans were grouped in two sets with vertical and horizontal views. The horizontal set showed slices of my head from the top of my skull to the start of my neck. The vertical set showed slices from left to right, starting with the left ear and ending with the right. It was weird, seeing that level of detail inside my own head. It looked like they'd sliced me open and peered inside. The clarity of the images was eerie.

There was a bright round pearl at the top of my neck and a shadowy mass behind my temple. The longer I looked at the pearl, the brighter it seemed. *What could it be?* When I looked at the top view of my brain, one side was markedly larger than the other. Was that what people meant by "left-brained"? It was time to call an expert.

"Bring the disc over here," Dave said, "and we'll look at it together." Feeling alarmed, I hopped in my car right away. Arriving at his house, we put the disc in his computer, and I pointed to the images in question. "That's not a pearl," he said, laughing. "It's blood. The MRI image slice shows it as bright white because it's a shot of fresh unsaturated blood heading up from your neck into the scanner's imaging field." We turned to the other image, with the strange-looking shadow. "That's your ear."

"See this?" He pointed to an image that showed the very substantial difference in size between the two hemispheres of my brain. "That shows that you didn't have your head straight in the machine. One side looks bigger than the other because the slices are tilted."

His lack of bedside manner had been honed by years of practice and thousands of scans. Listening to him, I wasn't sure if I should be relieved or disappointed.

After I got home, Cubby found the disc and turned the MRI data into an animated brain on YouTube. That was the most I could get, until the folks from the lab called me back.

"Everything looks good," Alvaro said. "When do you want to start?"

Returning to the lab a week later, I saw what they'd done with the MRI images. A large monitor sat on the tabletop beside the TMS machine. Its screen was divided into quarters, with the top left showing a circle with red crosshairs in the middle. The top right box showed my brain—all by itself in photo-realistic detail—suspended in space. Slightly to the right and below the brain I saw four red dots. "Those are the reference points— the bridge of your nose, the tip of your nose, and your tragus," Lindsay said, pointing to the small cartilaginous bit of my ear closest to my head. I reached up and touched that part of my ear. "Ear points," I said. She nodded.

"See those cameras?" Lindsay pointed at two cameras, which gazed down at me from the intersection of the wall and ceiling. "We use this"— she picked up a sort of laser pointer—"to highlight those spots on your head one at a time, so the computer can match the position of your head as you sit in the chair to the model on the screen."

The bottom boxes showed MRI slices of my brain from above and from the side. On each slice a set of red crosshairs was visible. "That's going to be our first target."

They fitted me with a headband that had three gray balls protruding from a two-inch-long shaft. "The cameras see the balls and the computer knows that the surrounding space contains your head. Now we'll do the calibration." She stepped over to the computer and started the calibration routine. The first point was my nose. Placing the dot on the tip of my nose, she looked over to see if the computer had picked it up. A soft beep signified success. Moving on, she touched a second point on my nose, then the tips of both tragi.

I remembered what I used to tell my son when he was little and needed a haircut. "Cubby," I'd say, "I'm going to take you to the barber and get your ears trimmed to nice sharp points like Mr. Spock on *Star Trek*." As a result, Cubby had resisted haircuts for years. Now they were tagging my own ear points!

With the system calibrated to my head they were ready to go. "We already calibrated the TMS coil before you arrived." Looking over, I saw it had a three-ball setup similar to the one on my headband. However, the pattern was different, which allowed the computer to distinguish the TMS coil from my head.

"Look," Lindsay said, and I turned to the monitor as she held the coil against my head. She moved it, and I felt it glide around my scalp. As she did that, the crosshairs in the three brain windows shifted, and a blue dot moved around the top left box, way outside the target crosshair ring. Then she placed the coil in position, and I watched the blue dot glide right into the center of the crosshairs.

"How precise is it?" I asked. In response, she moved the coil a tiny bit and I saw the dot move off the crosshairs. "It's got a resolution of about a millimeter," she explained. I was impressed.

Each of the MRI images—from both the vertical and horizontal series—was the same size and showed an area a bit bigger than my head in total. When you looked at the images you could see the skull, the scalp, and the faint outlines of my ears, nose, and mouth. They had gone into each image file and tagged the reference spots—points on my ears and the tip of my nose—one by one. That allowed a computer to put the data together and make a single three-dimensional model of my brain, in remarkable detail. All that was missing was my head. That was what I saw on the screen, marked with the target for the first stimulation.

It was a very sobering moment. Hearing "We're going to stimulate some regions in your brain's frontal lobe" as an abstract phrase was one thing. It was something else entirely to sit there in the lab, look at that image of my brain on the computer screen, and see that blue dot marking the first target. It all suddenly seemed very serious.

I don't think Lindsay and Shirley had any idea of the cauldron of feelings that were stewing in me at that moment. I don't even know if I could have described them. Wonder, curiosity . . . and a tiny bit of fear. But the thing that kept me going was hope.

"We can target any point on the outside of your brain, and this will show us where to place the coil," Shirley said with some pride. There was no question that they'd put a lot of work into modeling my brain. I hoped I wouldn't screw up and disappoint them when they put it to use. I seemed to have passed their tests so far. The MRI didn't show any missing pieces or extra stuff. Now that the mapping was done, things were about to get real. My first session was scheduled for the following week. I couldn't wait to begin.

The Night the
Music Came Alive

THE DAY OF MY FIRST TMS session started out like any other. I spent the morning at work, setting out for Boston in the afternoon. There was plenty of time to think during the drive out, and I tried to imagine what was about to happen and how my life might be different in just a few hours.

I was a mass of roiling emotions when I reached the hospital. The next step—anticlimactic in the extreme—was another series of tests in front of a computer, led by Shirley and her assistant. "You will see a series of faces flash across the screen. Just push the button that corresponds to the expression you see," Shirley told me. Left button for happy, right button for sad, with a third emotion in the middle. It sounded simple, but at the rate the images flew by, I had no clue what I was seeing. It was very frustrating, and I assumed there must be people who had the test down cold—otherwise, why do it? That realization made me sad; I felt like a failure before we had even begun, and my anxiety started gaining the upper hand over my excitement.

They gave me another test. This time Shirley put a microphone in front of me and started a new computer program. In this exercise I saw objects and had to say what they were, as quickly as I could. Dog . . . house . . . car . . . tweezers. The objects were all familiar, but the sequence made the words seem weird. Scissors, cat, and airplane did not normally go to-gether in conversation. This went on for quite a while, and I realized with a start that what they were testing was my ability to articulate what I saw. I wondered how changing that with TMS could possibly affect my life.

"Don't worry, that's just the baseline. You did fine," she said, as I sat with my tail between my legs. It all began to make sense—I'd do a test when I walked in, then they would zap me, and I'd do the test again. We all hoped I'd do better. For them it was academic. For me it was personal. Their tests and my beginner's anxiety had me convinced I was a failure. What if they booted me out of the study? That would be the ultimate hu-miliation, especially for something as trivial as failing to recognize birds and hammers quickly enough.

Finally we got down to it—the actual TMS. I'd experienced single pulses on my earlier visit, so I wasn't particularly nervous. But the ear-plugs and mouth guards they offered me this time were a surprise and I wondered if they signaled unpleasantness to come. I'd read that the TMS pulses could cause facial muscles to twitch, which I guessed was the rea-son for the mouth guard. Refusing it, I agreed to use the earplugs, because the fan that cooled the TMS machine was pretty loud.

Soon I was settled and the coil was in place. I thought of Alvaro's expla-nation of how TMS can be configured to enhance or inhibit the areas it is fired into, and although all the TMS they did was inhibitory, he'd also drawn that analogy of the balance between the two sides of the brain. Depressing the right would stimulate the left. So what would I feel? Ex-citement, depression, or something else?

Shirley stepped on the start button. The machine began humming, de-livering one pulse per second in a smooth, steady rhythm that would last for thirty minutes.

My proto-terror faded away with the first pulses. Nothing awful was happening. The pops reminded me of what I'd heard when we fired lasers, back when I worked at Candela. I felt a jolt of electricity that hit my head with every pulse—not enough to hurt me, but enough that I knew for absolute certain what it was. Anyone who's been shocked knows the feeling—it's unmistakable.

Is that electromagnetic induction? That thought passed through my head, but almost as quickly as it arrived, it was gone. All that remained was the march of time, as defined by the coil. *Pop. Pop. Pop.* My head twitched with every pulse, but oddly enough, it wasn't uncomfortable. It wasn't really anything at all.

Strangely, I felt . . . free. The roar of the cooling fan was loud in the beginning, but after a moment, the sound faded into the background. The continuing snap of the machine told me something was happening, though I couldn't feel it. In fact, I didn't feel anything other than my facial twitching. My anxiety had melted away, but what replaced it ran much deeper. It was as if I'd stopped thinking, and time seemed to stand still. My head was in a neutral sort of state; I had no desire to do anything but gaze idly at the clock on the wall. I wasn't even reflecting on my newfound tranquillity, even though it was a very unusual state of mind for me. As the clock's second hand moved in time to the pulses in my head, my world got smaller and smaller.

My ability to maintain a conscious stream of thought had slipped away, leaving me in a sort of mental standby.

More than once, I tried to count the pulses, but I didn't get far. One . . . two . . . three . . . four . . . and that was it. I didn't make a conscious decision to quit counting; the numbers just kind of ran down and stopped. My internal dialogue—the conversation that runs in our minds whenever we are awake—was somehow subdued.

The minutes passed and then, with one last pop, it was done. Half an hour had elapsed. I'd never have known but for the clock on the wall.

Looking around, I shook my head and slowly recovered my wits. Lindsay and Shirley were both watching me closely. A moment passed, and Shirley said, "Okay, let's do those exercises again!"

The list of objects wasn't as long as it had been the first time, and naming them was a snap. A moment later I realized there might be more than one possible answer to the things they showed me. Was it a dog, an animal, or a German shepherd? Was there one correct answer or many? Still, try as I might, I couldn't tell if I'd done better or worse than at the first session. As far as I could see, I was exactly the same as I'd been before, just a little bit stunned. What was I thinking? Had I really expected to sit down in that chair and stand up a different person?

Gradually, however, I began to sense that something *was* different. It took a while for the insight to settle in, because I felt as though I were moving through a dense mental fog. My thoughts were clear, but I had to be very deliberate in forming them, and I had to be careful uttering words and phrases. Not that I stumbled—I don't think I did—it was more that I felt I had to take more care, sort of like getting interviewed by the police when you've been drinking, or taking a mental walk barefoot over sharp gravel.

The scientists perhaps sensed that something was amiss. "Are you feeling okay?" Shirley asked, startling me. She repeated what she'd said before—she expected the immediate effects of this stimulation to last about fifteen minutes, and she watched me closely during that time. For what, I wasn't sure, because that time was lost to me. Gradually the mental fog faded away.

At the half-hour mark I thought I was back to normal. That meant it was time for the neurologist and his exit exam. This was something they had to do after every session, to ensure I was fit to return to the outside world. What was the alternative if they decided I was not? I pictured a locked cage in the basement. Perhaps it was one of those answers best left unlearned.

"What day is it?" he asked me, while watching discreetly for tics or twitches.

"Monday," I replied, with a touch of annoyance. Did they think I'd lost my mind?

And therein lay the problem, for it was actually Tuesday evening, as the neurologist revealed to me a moment later. "Huunh," I said in response. He looked at me, and I looked at him. I couldn't generally read the messages in other people's eyes back then, but I still watched for signs that he might be about to send me down to that basement ward for "observation." Then the moment passed. He let it go and asked me the date.

Son of a bitch, I thought. *This guy has hard questions tonight.* Venturing a guess, I said, "The eighth?" *I'm naturally ignorant about these things,* I thought. *It's not the TMS.* And I vowed to memorize the day and date before our next encounter. I didn't want to end up committed for observation over garden-variety ignorance.

"Where are we?"

"What county is this?"

"What floor are we on?"

"What season is it?"

The neurologist's questions seemed to fly past, and by the time we finished I was sure I'd been way more successful on his test than on the face recognition, calendar details notwithstanding. I wondered how the other patients had fared in comparison. Was I average, dumb, or exceptional? At least the neurologist had decided I was alert enough to be sent on my way. Once in my car I wove a path out of the hospital's parking garage and accelerated onto the highway for the long ride home.

The first call I made was to Dave, who'd been anxious to hear what had transpired. At that moment I didn't have much to report, beyond the mechanics of the experience. I was pleased to have made it out in one piece, unsure of what had happened, but seemingly okay. But as our conversation progressed I was struck by the realization that I sounded different.

Anyone who's sucked helium out of a birthday party balloon knows what it's like—there's no mistaking it. You take the sound of your own voice for granted, until it comes out sounding like someone else. The change here wasn't as drastic as helium breath, but it was noticeable, though I could not put my finger on exactly what was changed. I tried to unravel that mystery as I spoke, and gradually it hit me. Was it possible? My voice contained a tiny bit more emotional range, expressed as a rising or falling note to convey expression at the ends of sentences.

That was it. I realized I was speaking with more of what speech therapists call prosody in my sentences. But almost as soon as it hit me, the idea seemed crazy. *Could that be real?* I asked myself. *And what does it mean? Am I feeling more and expressing it in the same way or feeling the same and expressing it more? Or is it just my imagination . . . ?* I was so confused that I had to hang up the phone and be alone with my thoughts.

The scientists hadn't remarked on any change in my voice, and I wondered if they had noticed. But I forgot all about prosody a moment later when I plugged in my iPod and turned on some old music. That's when it hit me, like the onset of some hallucinogenic experience.

The first revelation came as I played the old Tavares recording, a live show from twenty-some years ago. It was an old bootleg, made off the mixing console during a small performance. It may not have had the sound quality of a studio production but those old songs brought back memories for me.

It's funny . . . I became known in the music world for creating wild effects for high-powered rock and rollers like KISS, but I was never a fan of that kind of music. I remember sitting backstage one night long ago with Peter Frampton and his longtime bass player, John Regan. Even though John had played with KISS's Ace Frehley and other rockers, he shared my feelings about heavy metal. "I like music that's more melodic," he told me, "with more sophisticated arrangements." I never forgot those words because I felt the same way.

Back in the seventies, soul groups were the epitome of smooth and melodic, and their onstage choreography was beautiful to behold. Of all the shows I worked, those were the ones I loved best.

I'd played the Tavares songs a thousand times before and heard nothing but an old bootleg tape. This time, everything was different. Every little nuance of the recording held meaning for me. My range of sonic comprehension had just widened a thousandfold. Whatever they did with that brain stimulator had unlocked something very powerful in the way I heard music.

The previous day I'd have heard a slight hiss in the recording and thought nothing of it. Now I could recognize the faint sounds of Chubby's microphone cable dragging as he walked across the stage. Each of the Tavares brothers sang in turn, and I smiled as I heard them sing together in beautiful counterpoint, always in perfect unison, even in the most complex harmonies. Thirty years ago—when listening was part of the job—I'd paid close attention to see if they were all on key and everyone sounded as they should. Nowadays I just enjoyed the songs and the memories they stirred. But I quickly realized something was profoundly different tonight. I was not just hearing more detail, I was feeling more too. All those years spent working in music, I had never felt I was sharing in the singer's emotions. Now, thanks to TMS, I was. Did other non-autistic people experience music in this way? I cannot know, but for me it was an unforgettable experience.

Perhaps this new way of hearing music meant even more to me because of my autism and my lifelong inability to feel what people say.

The Tavares set finished, and my memory followed the next song as it began. Things I'd forgotten a lifetime ago played back as if I were watching a movie.

In 1978, the Canadian songwriter Dan Hill had a number-one hit with "Sometimes When We Touch," and he used our sound equipment when he toured North America with Phoebe Snow. She was an even bigger star, launched on the success of her hit "Poetry Man." Now I found myself back on their stage, at the old Orpheum Theatre in Boston. Standing in the curtain folds, hidden on stage left, I watched Dan Hill out there alone, silhouetted by the spotlights. The melodies he played were crystal clear, and I remembered every detail of our setup that night. The Orpheum, Capitol Theatre, Vassar College . . . I'd carried their gear all over the Northeast, filled with pride that my amplifiers were delivering their music to the crowds.

As the next song played, the musicians and the venue changed and I heard Diana Ross hit a triangle. As it rang out I saw her in my mind, standing at the front of a different stage, holding the chime up to the microphone. The metal rang with a beautiful purity, and I felt the joy and energy in her voice. She wore the most dazzling sequin dresses at those shows.

My mind wandered to other times, when Diana was dating Gene Simmons, the bass player for KISS. She would stand silently beside me backstage as we watched him play from the shelter of the speaker cabinets. As we stood there I would see the waves of music, as if there were an oscilloscope inside my head. The songs themselves had become tangible *things*. I could reach into them and hold the individual bits and pieces— melody, rhythm, instrumentals, vocals—in the hands of my imagination.

All that came back to me now. It was like a replay of some of the most pivotal moments of my young adulthood, with a brand-new layer of emotion laid on top.

I heard Eddie Holman announce his song "Hey There Lonely Girl," and I could see him up onstage in some long-lost arena. At the end, I felt his joy when he shouted, "Thank you, Lord!"

As the guys in the band talked between songs, I thought of my friends— Bobby Hartsfield and Seabreeze, the brother of blues musician Taj Mahal—standing outside my shop with their Harley-Davidson motorcycles. As I listened to the recording of the musicians, my old friends, I could see their individual expressions and understand what they were feeling. Their voices had a beautiful richness and warmth. I had always heard it but now I could feel it too.

The filter of autistic disability—if that is what hid the emotion from me before—seemed to have vanished. I heard a smile in one voice, as I saw it on my friend's face, and I felt its truth inside of me. All the while, the sweet music kept on playing. *If only this could last forever,* I thought. Then I remembered the fifteen minutes Shirley said the TMS effects would last. We were way beyond that now. What was going on? Was this the real goal of their research or some bizarre side effect? Or worse, was it my overactive imagination? Either way, my insight into what I heard felt incredibly real.

When I listened to McFadden & Whitehead sing a song they wrote for Marvin Gaye, I focused my mind into the performance and listened to the instruments one by one, as I'd done so long ago when I was a music engineer. What had been a background melody resolved itself into notes from a keyboard. As I concentrated, I realized I was hearing several keyboards, each with its own distinct sound. The more I focused, the more clearly I recognized the individual instruments and their arrangement on the stage. There were three stacked keyboards with a fourth—a piano—

to the side. Their sounds were so clear; I felt I could touch them. Just then the keyboard player was walking one hand along the Korg synthesizer while his other hand played the Hammond organ. The melody and counterpoint made me smile as I admired the masterful way he played the difficult passage.

Was I remembering shows long past, or was I hearing and recognizing the individual instruments in a new way and then building the images in my mind? I don't know. But I trust my ability to recognize different instruments when I hear them, and the subtle sonic cues were enough to tell me where each performer stood up there on the stage. Back in my twenties, I could not only tell a Gibson bass from a Fender bass, I could tell one Fender from another, and even what kind of strings they had. I couldn't experience the emotions in the music, but I could hear all the parts and understand how the music was made.

Many symphony musicians say they can do the same thing. Most people don't hear with that level of detail, but it was that precision that had taken me to the top of the music-engineering world. As that thought entered my mind, I realized that the abilities I'd left behind had all come back to me, with added emotion.

In *Look Me in the Eye* I'd written dispassionately of losing that gift, telling readers it was a good trade-off and part of growing up. And as I lost the ability to see deeply into music, I moved away from that scene. In exchange I believed that I'd developed an ability to relate to people, have conversations, and make friends, and I'd largely forgotten what that sense had been like. Now that it was back, I was overcome with emotion. After all those years I suddenly—and profoundly—valued what I had lost, and I treasured its return.

And it was totally unexpected.

The moment I got off the highway I called Bob and Celeste. I was so overcome by the power of the music and how it came back to me that I wanted to tell them all about it. "You have to meet me," I said urgently. We agreed to rendezvous at a restaurant in Amherst, where I tried to explain the experience, but I could not stop crying as I struggled to put it into words.

Once I got home I played more of my old recordings long into the night. Paul Stanley roared the introduction to "Shout It Out Loud" for KISS, and Melissa Manchester sang "Just Too Many People" at a long-ago concert by Cape Cod Bay. I continued to smile through my tears as the emotions of the songs washed over me like a warm summer rain. But as the night wore on, the magical edge started to fade. The researchers had told me that the TMS effects would only be temporary, but I so wished for this wonderful power to last forever. By five A.M., the brilliance was gone, and I fell asleep.

My family slept through the whole thing. When I woke up the next day, my hearing seemed to have returned to normal. The crystal clarity of heightened perception I'd had the previous night was gone. Sound was no longer a portal into the hidden workings of everything around me, as it had been for those few hours the day before. The extra range I'd sensed in my own conversational voice seemed to have slipped away too. Yet I remained incredibly moved by the experience. My "normal" wasn't quite the same as it had been two days earlier. I felt a mixture of sadness and wonder at the way it had all unfolded.

Lindsay, Alvaro, and Shirley didn't say much in response to my email describing what had happened. All Alvaro wrote back was "Very interesting. And unexpected." When I asked for more, Alvaro said they didn't have any idea why the TMS would have awakened the musical vision. It hadn't been part of their plan.

I began to suspect that they did not know how these experiments stimulating the mind would play out long-term either. Before we began I'd imagined that they had done what they were going to do to me many times before, so I was testing a familiar thing in a new context. Now I was discovering that wasn't the case. If what they said was correct, the result I experienced was not only unexpected but a mystery.

The next day I called Dave and filled him in too. After listening to my story, he again offered up his physician's perspective. "They had an idea, a treatment plan, and a test to see if their idea panned out. You did that, and all this stuff you just told me about came later. So the real experience for you was from the side effects. You don't even know if their treatment did what they hoped, because no one told you what they intended. They couldn't have expected what happened later, because if they had, they would have kept you at the hospital to measure it. They sent you home with no idea what would happen later." I realized he was right, and the idea left me more than a little unsettled. "I think it's kind of neat," he told me.

Before we hung up, I assured him that I was still developing amazing superpowers, and even then I could practically visualize the numbers on his credit card. And that was with one session! By the third session, I would surely be able to clean out his bank account, just by thinking about it. He expressed his doubt and I said, "You just wait and see."

Later that day I was certain that all the effects had dissipated, and I felt a bit sad. But over the weeks that followed I came to realize that my initial judgment was premature. When I listen to music now, it does not have the mind-blowing richness of what I heard that night after TMS. But it still feels fuller and more detailed than it had seemed in years. Little details like the brush of a cymbal would have escaped my notice before TMS. Now they stand out clearly. Did TMS make me notice what was already there, or did it help me see what I'd been blind to for many years? I don't know, but whichever it is, it's wonderful.

And a remnant of the emotional change seemed to linger too, making my listening deeper and richer, more full of feelings. Not only do I recognize a wider range of musical sound, I also associate what I hear with a broad range of feeling. Some of that is still with me today. The best way I can describe it is with an analogy.

Imagine that all your life you have seen the world in black and white. Meanwhile, everyone around you describes the beauty and richness of color. After a while, their talk of color frustrates you. Which do you believe? Their words or the evidence before your eyes?

You say to yourself, *This color thing is bullshit. They see what I see, and they say "color" to describe the same shades of gray. They're putting me on!* Now try to imagine what it would feel like to experience a glimpse of the truth. You step into a lab, and for a few hours, scientists turn on your ability to see the world in all its vivid color. You realize that what people were saying all along was true. It was your senses that were deceiving you.

Then the colors fade. Your world is once again black and white. Yet you are forever changed. Before, "color" was just a word. Now, it's a memory, and a vivid, compelling one at that. You strive to make it real, and you interpret everything you see in light of that remembered color.

One day, perhaps the colors will return for good. Until then, your ability to imagine them remains, and your world is transformed by the memory, even if you never see beyond black and white again.

When I described the experience to my friends later on, some of them asked if it made me feel cheated—gaining insight only to see it slip away. I just smiled. I'm not a churchgoing person, but for me the experience felt miraculous, like a religious vision. For a brief time, I felt a deeper reality, and even as it was happening I knew my memory of the experience would never leave me.

Alvaro and I discussed the color-blindness analogy a while later in his lab. My experience may have been an unforeseen side effect, but he was ready and willing to explore where it led. "If we use color blindness as an example, you grew up having to explain away what everyone around you said about color and emotion. Because we tend to believe our own senses, you decided incorrectly that everyone else was wrong, and even worse, they were putting you on. That's an example of what we call maladaptive behavior.

"So you would listen to people talking, and they would be expressing emotion in their words, but all you heard was the logical meaning and you eventually started to get angry when you failed to understand what they really meant.

"To me, what the TMS session illustrated was that your ability to see sound was never truly gone. You made paths in your mind when you were young, and then you went on to other things and lost the ability to find them again. But they were in there all the time. Somehow, TMS showed you the way back. Perhaps TMS took them to a greater level, temporarily. The mind is a complex thing."

"Will the brilliance of music return again?" I asked.

He looked thoughtful for a moment, and I sat there, silent. In the end, "I don't know" was all he could say.

Over the next few days I thought about our conversation and what I was feeling. At first I'd thought the value of the experience lay in the momentary return of my ability to see really deeply into music. But gradually I came to understand that there was something more—the partial restoration of my "old" music vision was now tied to a new layer of emotional understanding that I'd never had before. And that's what would remain, long after the crystal clarity of that unforgettable night's "musical hearing" faded.

Emotion

AS THE DAYS PASSED, I came to see that the color-blindness analogy was even more appropriate than I'd realized at first. Knowing there was a whole world of emotion hidden within each song made me hear music in a different way, and this remains true even as I write this, seven years after my first experience. Though what I hear and feel today is nothing close to what happened in the car that night, I still have emotions I never felt prior to TMS. I may have listened to a particular song a thousand times over the years with no effect, but after TMS I feel its meaning, sometimes very intensely.

The emotion that made me cry in the car was not sadness. The best way to describe it was intense, or even a little bit scary. I used to listen to music and admire the technical excellence of its production, or criticize the recording's errors. I might like the flow of the melody or the rhyme or meaning of the words. But the emotional message of a song had never meant much to me. Now I found myself bursting into tears when I heard a beautiful song.

For a person who'd always been as logical as Mr. Spock, this was a very unsettling state of affairs, one I tried at first to understand rationally. I couldn't talk to my friends about it, because I felt like a fool, getting all teary listening to music. All I could do was keep quiet and ponder.

I thought of composers and performers I'd known over the years and how they often interpreted the songs they wrote or sang in differing ways. Songwriter Jimmy Webb said he'd imagined "Galveston" as an antiwar song, but when Glen Campbell made it a huge hit he called it a "march to war." That made me wonder if everyone shared the feelings I was having in response to music, or if different people responded in different ways. Might I interpret a song as happy while you understood it as sad? As I tried to define exactly what I was feeling, I realized I couldn't even describe in words some of what was welling up inside of me as I listened. It was just . . . raw emotion. Not happy, not sad. Just strong. Eastern spiritualists call that Kundalini energy.

If that wasn't enough, I began getting overcome by things that I read. I'd be at the breakfast table, reading a story in *The New York Times,* and I'd be blindsided by emotion and have to stop reading. These new feelings were very strange, because they were incredibly strong and they seemed to be triggered by events—in songs or news stories—that had never before elicited the least bit of response from me. In fact, in the past, I had belittled people who burst into tears at the news of a bus crash at the other end of the world. "They don't know anyone on that bus," I'd said dismissively. "They don't even know anyone in that country. It's just a play for attention." But now it was happening to me.

My first reaction was surprise. Then I felt ashamed when it started occurring in the presence of other people. There was no real reason for me to be upset by news stories about strangers, but my feelings were unmistakable. It was enough to put me off newspapers.

Martha would look at me quizzically and ask what was wrong, but I shrugged it off as if nothing had happened. This new emotionality was not something I was ready to talk about. She was already very worried that TMS would change me, and even though I didn't understand why, I sensed talking about it would make things worse. "We have so many shared memories and experiences," I told her. "TMS isn't going to change those or take them away." Yet my words didn't soothe her, and I wondered if they were mere wishful thinking on my part. "You don't know what will happen," she said. But I couldn't see any danger in the experiments and I was entranced by the experience.

I thought I might seek counsel from Alvaro and the others in the lab, but when I tried to describe the way I felt, I found myself at a loss for words. Usually I had words for everything because I'm a very rational, logical guy. But now the naked emotion stood alone. "I feel things, but don't know what they are," was all I could say.

To put my situation in perspective, I recalled a moment a few years earlier when my father was sick and in the hospital dying. As I looked at him, a little voice inside me said, *He's going to die.* A feeling of terrible grief washed over me and I started to cry. The few times that had happened before in my life, the voice and the feeling had always been intertwined, so I understood why I felt as I did. Now they weren't. These strong emotions I was now experiencing were unaccompanied by any insight into why I was feeling them.

Reading or listening to music around other people now made me self-conscious, because I never knew when some innocuous passage would leave me inexplicably teary eyed. As an author at speaking events, I used to read aloud in a clear, calm, steady voice. Now it seemed like anything could set me off and my oratory would fall apart. I barely trusted myself to speak in public.

The research is funded by Robert Wilkins, who gave the medical school sixteen million dollars after the death of his son Alfred. Find a cure, he told us, and that's what we hope to do.

How hard should it be to recite something like that? Every medical school in the country has passages like that in its literature. Reading them was now impossible for me. I would be okay until I got to the "death of his son Alfred," which hit me like a punch in the gut, and I could hear my voice waver when I read, "Find a cure." Later, when I would recover my composure, I'd reflect on how crazy that seemed. Mr. Wilkins was a stranger to me, and I'd never heard of his son. Hundreds of thousands of people die every day. Why should those words affect me at all?

Melody affected me the same way. At first I thought it was the words of songs that were hitting me so hard, but I listened to classical compositions with no vocals and felt the same powerful emotions well in me with the ebb and flow of the symphony. When I experienced that on the ride home from the hospital it was miraculous: new and wonderful. A week later, it was scary—like a toothpick castle built without glue. With the slightest touch, I'd fall to pieces.

Could one TMS experience have done all this? I wondered.

The researchers had promised that the effects of TMS would be temporary. They had all said it more than once. *This can't be happening,* my logical mind reassured me. *TMS energy has to dissipate over time. It's not self-sustaining. You're imagining the whole thing.* But as the days passed, my sensitivity to emotions seemed to get steadily stronger.

"What do you think is happening to turn on these emotions in me?" That was my question to Alvaro on my next visit to the lab. One of many great things about Alvaro was that he never ridiculed my thoughts or feelings, no matter how strange they may have seemed. Whenever I asked him a question, I always got a reasoned answer, to the best of his ability.

What he said was very interesting. "I think there's a mechanism in our brains that helps us see expressions and body language in other people and act out those things in our own minds, so we can feel them ourselves. It responds to inputs from all our senses, even sound and smell. Your mirroring system has to have a regulatory mechanism, otherwise you would be overwhelmed by emotion all the time. We have hypothesized several regions where that regulatory system might be physically located in the frontal lobe, and we are now going to suppress those possible targets one by one with TMS.

"Our theory is that all people have this wiring, but in autistic people the regulatory system is overactive, preventing the emotional messages from getting through." We had talked about this when we first met, but now I was matching Alvaro's words with my experience, and they fit remarkably well.

"I see. . . . So when you suppress that network in my head, and I start to use those paths, I build connections that keep going even when the TMS wears off. Is that why I'm still feeling these things, even after the TMS effect went away?"

"Exactly," he answered. "At least, that's what we hope. And the fragility you feel may be because it's new to you, and you have not yet learned to adapt. If that's happened, it's great. But we have to be careful, because when things work out better than you expect, there may be a surprise coming. So we have to watch and see."

His choice of words—and the possibility of a surprise—was a little worrisome, and I said so.

"Everything you reported so far has been a surprise," he responded with a smile. "We hoped for good results, but what they would be was not exactly clear." His caution seemed reasonable, and his notion that I was emotionally vulnerable because the feelings were new made sense too. Maybe this was like getting a filling at the dentist. Any newly repaired tooth is sensitive to hot and cold for a week or so until it acclimates. Lindsay was in an office down the hall, and I wandered down to see if she agreed. She must have better teeth than I do, because she wasn't sure about the dentist part, but she did agree on the neuroscience. Mirroring networks were actually the subject of her doctoral dissertation back at UCSD, and her work had been published to considerable acclaim in 2005 and 2006. "I'm sure it helped me get the job in Alvaro's lab," she told me later with some pride.

Could what I was feeling truly be called mirroring? I pondered the idea. To me, mirroring implied an immediate response. There's no delay when you look in a real mirror. But the new emotions I was feeling were somewhat delayed. I was hearing things and then having feelings well up as long as a minute or two later. Also, the word "mirroring" didn't seem to be the best choice to describe a response to things I heard or read. Mirroring implies sight, or mimicking the emotions of another person, but I was now responding to words on a printed page, not a person speaking in front of me.

Autistic people are often excessively literal, as I reminded myself mid-thought: *We are trying to explain what's happening here, not pick the per-*

fect word. Then I thought back to a section I remembered reading in one of Lindsay's articles: "Are mirror neurons involved in the ability to understand metaphors? Autistic individuals typically have difficulties with metaphors, often interpreting them literally, and the researchers believe this too may be connected to a dysfunctional mirror neuron system."

That was exactly what was happening to me. What was music if not melody and metaphor? And if that was so, my understanding of music was surely enhanced by TMS. Mirror neurons might or might not lie at the root of the change. Whatever the explanation turned out to be, I knew that Lindsay, Shirley, Alvaro, and company were onto something big.

As quickly as I had that thought, I felt the beginning of tears welling up in my eyes. *Tears of joy?* I thought. *Tears of excitement? Tears of confusion?* Once again I didn't quite know how to describe my feelings. My systems were overloaded, and I could no longer tell up from down, emotionally speaking. TMS had surely started me on a journey, with no predicting where I was headed next. For a person who'd been ruled by logic for fifty years, this new topsy-turvy way of experiencing the world was quite a change.

Singing for Ambulances

A WEEK LATER, I went back to the lab for two more stimulations. As Shirley explained, "We hoped the first stimulation would raise something up in you, and we hope the second will lower it back down." But she didn't tell me what exactly would be raised or how it might affect me. So far I'd heard talk of raising and lowering, making new connections in my brain, and opening up suppressed pathways. I assumed those were just analogies to convey complex science to me, but the changing explanations actually left me more puzzled and confused. Now Shirley's enigmatic plan for the day left me a little bit uneasy, particularly given my last experience. The musical vision that followed the previous stimulation was remarkable, but the emotional instability I was feeling now was difficult to manage.

I wondered what would happen this time.

This day's stimulation would target a new area of my brain, farther forward, between my right eye and ear. My wife accompanied me on this visit, and she watched me through the process. I had the same feelings of stopping time or being in a meditative trance, but there was also some discomfort because the TMS energy was flexing nearby muscles on my face. Afterward, Martha said my face had twisted with every pop of the coil, but the thing that most disturbed her was my strange expression. "You looked like you were smirking over some joke none of us was in on." I was surprised, because I hadn't experienced it that way at all, and there certainly wasn't anything funny about the TMS.

Her words gave me an odd feeling. She had always been good at watching me and the people around me. Ever since we'd learned I was autistic, she had tried to act as my emotional eyes and ears, and I relied on her for that. Now she was describing emotions on my own face that I did not remember feeling. It was one thing to have her say "that person is not sincere" about a fellow I was talking to. It was something else entirely to have her tell me there were unexpected dramas playing out on my own face that I had no awareness of at all.

The contrast between her observation and my own recollection made me question what was really going on. With such a gap between my remembered experience and her observations, how reliable an observer was I?

My nerves were all jumpy, and I felt rattled as I walked out of the lab. That state of agitation persisted as I did the follow-up tests on the computer. I felt like the questions were jumping out at me, even though the computer was silent and the test itself was exactly like ones I'd done before. When I thought about how that made me feel, what came to mind were accounts of people using psychedelic drugs in the 1960s. I'd read about words jumping off a page and grabbing you by the eyeballs, and at the time I'd thought it sounded bizarre and funny. Now, to experience that firsthand, with no drugs in my system, was very peculiar indeed.

Could TMS energy change my mood? That was a good question, and I immediately remembered the work Alvaro was doing with depression. *Of course it can,* I told myself, and my experience today was the opposite of what they must hope for in treating depression. When I saw Alvaro again I asked him about that. His answer surprised me.

"Responses like that are a funny thing. If we do the depression protocol on someone who isn't depressed, it can have the opposite effect of what we intend. It can make him unhappy and anxious. But that is a different brain region than what we are stimulating in you, and the TMS pattern is different too." The incredible complexity of the mind was becoming more and more apparent to me with every step of the process. The limits of knowledge of even these experts was unsettling, just as the way they were pushing the envelope was exciting. At least that was how I saw it. Unfortunately, those thoughts didn't make me feel any less rattled.

As I wrote to Alvaro in an email later that evening, "If I were to choose a single word to characterize my feelings right now, it would be: jarred. I don't know if it's the TMS energy in my brain that leaves me feeling this way, or if it's a result of the 1,800 contractions of my right lower jaw muscle.

"It's almost as if I'm irritated at something, but I'm not."

Back in the lab that afternoon, Shirley seemed to see that, and she suggested some time off to let my brain settle down. "We need to take a break between the first and second stimulations," Shirley had explained earlier. "Why don't you walk around, relax, get a bite to eat. Let's meet back here in two hours, okay?"

Martha and I went downstairs to an open-air Starbucks and sat outside sipping tea. I ate a cookie. My head was still roiling from my time in the lab, and I sat quietly, watching the activity around us. The scene seemed pretty normal at first. Then an ambulance pulled up alongside the curb— maybe twenty feet away—and briefly turned on its siren. Barely a beat passed before I opened my mouth and howled right back at it, at full volume. Martha looked at me strangely, and I wondered why. At that moment, singing along with the ambulance seemed like a completely natural thing to do, and I gave a fine, melodic cry if I do say so myself. The others in the café didn't seem to agree, but fortunately, no one called security. I had learned what I thought was a likely explanation for that a few weeks earlier, from Kathy Dyer, a speech pathology professor in the autism program back at the Elms. We were talking about how my autistic eccentricity was perceived, first while I was a child and now as an adult.

"There's this thing called the competence-deviance hypothesis," she explained. "It says that the more competent an individual is in his field—the more respected he is in the community—the more his eccentric behavior will be tolerated by others.

"But the opposite is true for young people, because they have not done anything to earn respect in a community. So when they do weird things they are treated like dangerous animals and hustled into cages. It seems unfair when older, respectable members of the community do stuff that's even stranger and people just shake their heads and smile at their eccentricity." I'd been aware of the need to appear responsible and respectable ever since my book came out, so I was wearing a nice button-down shirt and I was clean-shaven and showered to boot. The way I looked, I might have been a doctor or Harvard professor. If I'd been sporting a scruffy beard, a leather biker jacket, and a motorcycle chain around my neck, the patrons might have reacted a bit differently.

At least that's how I explained what had happened to Alvaro and Shirley a couple of hours later when Martha told them the story. In the short time that had passed, the whole incident had faded from my memory, and I wouldn't even have thought to mention it to the researchers if she had not brought it up. "You might not have made much of it, but I'm sure everyone else in Starbucks still remembers," Martha assured me. "The whole place was looking at you—a great big guy with his head thrown back, howling along with the ambulance. You were like a wolf, singing for his pack."

With a bit of trepidation I smiled at her description, but then I realized it wasn't a compliment. I felt that cringe coming on—the one I get when I realize that I've really screwed up and it might be too late to do anything about it. Alvaro and Shirley didn't say much in response. The strangest part was, the howl had meant no more to me than an automatic "excuse me" when making my way out of an elevator. That's why I hadn't remembered it. Who knows what I might have done had Martha not been there to look askance at me.

Was that an effect of the raising up Shirley had referred to, when she described the day's two sessions? That sounded like an appealing explanation, because it meant that something in the TMS had caused the howl, and I wasn't just crazy.

Shirley's response was to do more TMS, in a nearby area. This time I had to clench my jaw to keep my teeth from clattering together, which was somewhat uncomfortable. The tests I did before and after the stimulations explored prosody—the way my voice may have changed in response to stimulation.

I repeated phrases like:

"Mike lives at Thirty-four Alford Street, in a green house."

"John works with roses, in the O'Connor greenhouse."

Supposedly, the lilt in our voices helps tell listeners the difference between a home that's painted green and a glass-sided structure filled with plants. Did I do better or worse on the tests? I couldn't tell. As we drove home, Martha paid close attention to how I sounded. She thought my voice had a bit more tonal range, and I agreed. It wasn't quite the same as last time, though. Before, I had sensed a lift in pitch at the end of my sentences. This time, it felt like the change in tone extended throughout the entire sentence. My son noticed it too. When we got home, he said, "Your whole voice goes higher pitched now and then drops back to normal."

It felt a little strange listening to their comments, as if I were a rat in a cage and they were observing my changes as the scientists experimented upon me. It was also interesting that I could comment on my seemingly increased tonal range, yet I wasn't being overwhelmed by emotion anymore. What would Cubby have said had he been there for the ambulance howl? Luckily, I never spontaneously howled again, or if I did, no one else was there to point it out. That night I lay in bed pondering the day's events—particularly singing along with the ambulance. *Why did I do that, and what might it mean?* I felt like a kid again, searching for a way to explain my outlandish behavior to the adults around me.

Your sense of inhibition must have been suppressed. That's what the voice in my head told me, and it may have been right. I wondered if my singing out was an example of the mirroring I'd been learning about. Resolving to look into that the next morning, I drifted off to sleep.

The next day, some additional reading about mirroring and mirror neurons made clear how far off base my ideas had been, at least with respect to published science. Mirroring in human beings begins when an infant looks at Mom's face, reads the signals, and smiles back, learning to feel Mom's happiness. Normal humans do not mirror machines— ambulances or otherwise. *You're not normal,* the voice in my mind reminded me. I wondered if that voice was right. Maybe mirroring was different in me, because I'm autistic. I sing for ambulances.

As nutty as it sounded, I thought there might be something to it. After it was first observed in monkeys, mirroring was then found in a bunch of other animals and, of course, in humans. Neuroscientists have identified the particular brain cells that do this job, and there is a lot of debate over the way they may be implicated in autism. Some researchers think mirror neurons are missing in autistic people, while others suggest they are broken. And then there is a third contingent of scientists who think the whole mirroring theory is wrong and that autistic social blindness has some other root cause.

It's a frustrating thing to try to understand, because the experts have such disparate and mutually exclusive views. There are a few scientists who assume autistic people lack empathy and emotion completely. That sure doesn't describe me! When I think back to how lonely I felt as a boy, it's hard to believe that any person could feel more sad or more alone. I'm not sure the researchers who hold this belief really know what they're talking about or know autistic people. I have a hard time believing that I am the exception to the rule and that other autistic people are emotionless automatons.

Then there are the scientists—and this seems to be the majority—who feel that autistic people have the same emotions as everyone else but we don't have the expected empathy reactions to things that happen around us. That's true in my own life, and I've seen it in others' too.

When I get upset, I'm sure my distress is as intense as yours or anyone else's. Now imagine that you and I are crossing the street, and you trip and scrape your knee. A non-autistic person might "feel" your pain, even though nothing has hurt her. Even though I don't always share your pain, I'm fully aware that you fell, and I'm ready to respond. A typical person might offer a sympathetic kind of sound, while I'm more likely to offer a truly practical response, like "Get up!" That may sound cold and uncaring, but it's not. The best anyone could do is to get you out of the street before a car runs you over.

So I might not have the expected response when you trip and fall, but you could call me on the phone and tell me about some bad medical test results, and I might get more emotional than anyone else you know. When that happens, anyone can see that the ability to have the feelings is there inside me, it's just that my reactions are not triggered in the same ways. And from what I can see, lots of other autistic people are the same. The question is why it happens.

When I asked Lindsay, she said, "That was the subject of my master's thesis. We studied the nervous system reaction by measuring how much kids sweated when they saw someone else get hurt. That's one measure of empathy. When we compared autistic and typically developing children, they had equal physiological responses—all the kids sweated the same amount in response to stress—but their visible behavioral responses were strikingly different.

"So their nervous systems reacted the same way, but you couldn't tell by watching them. There was no visible sign that the autistic kids felt anything at all, even though their clammy skin said they did.

"I proposed that the connection between the visual and emotional centers was intact, but that there was some breakdown between the emotional centers and frontal regions of the brain that help produce a behavioral response to what you see. So the autistic kids felt the same things but didn't show it. That was how I came upon mirror neurons and hypothesized that they may be the key to this breakdown in empathy.

"And by the way," she added, "no one was really hurt in the study. We showed the kids video clips of actors pretending to poke themselves with needles, and they responded just like it was real."

I understood Lindsay's explanation, but the more I read about what she had called internal wiring differences in folks with autism, the less I liked what I saw. Most descriptions seemed devoid of hope for change, and it was hard for me to accept that there was nothing to be done. But my best course of action was to think positively. What if my mirroring system wasn't hopelessly broken? Alvaro had suggested that, and Lindsay was optimistic that TMS could help change my wiring. Perhaps my mirror system worked but mirrored something besides faces. Who's to say that the only kind of mirroring is between two humans? With my close connection to electronics and machines, wouldn't it make sense that I might watch a machine or device and imitate it in my mind? People always said I had extraordinary insight into machines. Maybe I was using brain areas that most folks use to understand Mom to understand a Jaguar V12 or Fender Twin guitar amplifier.

When I worked in music, I always felt as if I sensed the emotions of the equipment while everyone else felt the emotions of the songs. It's strange to ascribe feelings to pieces of electronic machinery, but that's the only way I can describe the way they called out to me. For example, our sound system had one hundred–plus speakers, divided into groups for different parts of the sound. The biggest speakers carried the heavy bass notes, while the little compression drivers and horns delivered the highest highs—the brush of the cymbals, the snap of a snare drum, and the brilliant edge of the other instruments. Bass drivers tend to be big, slow, and forgiving. Horns are not. The sound of a horn's compression driver changes dramatically when the amplifier behind it overloads and goes into clipping. With one hundred speakers in the system, it's hard to recognize each one separately and pick out those cries, but I taught myself to do it.

What I'd hear was the horn squealing in pain. Ignore it, and the horn's diaphragm would shatter. The grunts of a straining bass speaker are a bit more subtle, but the consequence of ignoring them is just as destructive. Cones rip, and amplifiers blow up. So my job was to get the best sound and clearest volume without any destruction. The audience never noticed, because I felt that pain the moment anything bad occurred, and I backed down the crossovers and turned up the limiters to protect them.

Sometimes people on the crew would remark on my ability to do that—"It's like a part of you crawls right into the wires, like an alien in a movie"—and they'd suggest that I was part machine too. At the time, I knew nothing of autism and I assumed any sound guy could concentrate the same way. Now I know most can't, and I see how my ability was both a gift and a difference that set me apart from everyone else.

Could another autistic person be just as likely to deploy that mirroring system to see into the hearts and minds of animals and, in doing so, become an extraordinary animalist? If such a thing were possible, that would truly be a situation where one man's disability was another's great gift. I liked that idea a lot. It made me think of my friend Temple Grandin, an autistic woman ten years my senior who has written very successful books about her experiences and who often describes seeing the world the way cows and other animals see it.

Unfortunately, Alvaro didn't agree with my theory. "I don't think that's correct," he said when I presented him with that notion a few days later. But there was a hint of doubt in his voice. The question had gotten him thinking.

Meanwhile, there didn't seem to be any new long-lasting effects from this most recent round of TMS. I kept waiting for the other shoe to drop, but nothing happened. That made me curious to know whether the effect of the last stimulation was too subtle to be seen or if it would show itself later. "Some of the areas may have no effect," Shirley had told me before we began. With the passage of time, my emotional fragility was lessening, but I still felt all sorts of new emotions from things I read and heard.

Cubby and Martha thought my voice had returned to normal, though I hoped they were wrong. The problem with watching for subtle differences of the sort I was experiencing is that we become accustomed to a new normal very quickly, and that blinds us to little alterations.

Both of them had been quick to say, "You sound different," but determining when and if my voice returned to normal was quite a bit harder. Interestingly, I wasn't reading body language or expressions any better—at least not that I could tell. Maybe that meant the scientists had missed the mark with me, but I kept my mouth shut because the other changes I was seeing were nothing short of remarkable. Even in my self-centered and oblivious state, I knew better than to insult them. They had done these experiments to see if I would change, and I had. The only question was where it would lead.

By this time, there were several different TMS studies going on in Alvaro's lab, with teenage and adult volunteers. But when I asked the researchers about them, all I got was that enigmatic smile. "We can't tell you about the others," Shirley would say in her French accent.

A Family Affair

MY FAMILY'S REACTIONS to the way I was changing ran the gamut. Martha was worried. Little Bear remained skeptical. My mother was fascinated. And Cubby's response morphed from indifference into curiosity. Now he wanted to get involved and decided to join the program as a research subject.

Their differing reactions reflected the personalities of the members of our little family. Martha was depressed, so she tended to focus on the potential pitfalls in any situation. Despite that, she'd always trusted me to forge ahead. But that didn't seem to apply to the present situation. TMS seemed to make her nervous no matter what I said. "You hope it turns out great," she told me, "but you really don't know."

Like many ex-spouses, Little Bear and I had a somewhat combative relationship. We'd shared many interests when we were younger, and we still had a lot in common. It hadn't seemed that way when we first got divorced, but I'd come to see it with the passage of time. Our greatest interest of all was our son, Cubby. Yet our once-intertwined lives had become quite divergent, and the degree to which we'd grown apart was at the root of our divorce. Both of us had escaped violent, crazy parents—

and we'd seen the inability of psychiatry to fix our childhood families. That made us both dubious of any possibility of changing the brain. She had watched the failure of psychiatry with my parents, and she had shared my terror when they were each institutionalized during junior high school. Rightly or wrongly, she saw TMS as "psychiatry on steroids."

But I was on the inside, and the changes I felt seemed more real than any I'd known before. I told that to my ex-wife, but she was not ready to accept it. Try as we might, neither of us fully understood where the other was at, and she greeted any news of mental experimentation with great skepticism.

My son wasn't scared or skeptical. He was just in his own world, one where chemistry and physics were front and center. I'd tell him what was happening in the brain lab, and he'd respond with his latest discoveries in science. With their collective absence of enthusiasm I realized I was on my own with this TMS business. I'd done other things alone—like genealogy research—and they had worked out. I trusted myself to stay the course. Just four months had passed since that night Lindsay approached me at Elms College. It felt like much longer, so much had changed. My family didn't really share my feeling of incipient transformation, but they did see that something was happening.

That was what ultimately drew my son into the study, and I was really happy to have him join me. We had always known Cubby was a lot like me, and that was confirmed in a sort of elemental way when he was officially diagnosed with Asperger's that spring. With his new diagnosis, Cubby qualified to join the study, and he did so at the end of April. He'd gotten into some trouble two months before, and the TMS study seemed like a good way for him to learn about himself while providing a distraction from his current lot in life. It turned out that my son had felt the same sense of social isolation that I had known, even though he was far more successful socially. I hadn't really grasped that until we spoke with the psychologist together at the time of his diagnosis. Like me, he was willing to try anything that might help him connect a bit better, even though his main focus remained his science experiments.

Cubby's fascination with chemistry started out with model rocketry, which led to making rocket engines, which led to the pure chemistry of energetic materials—explosives. He would mix up experimental compounds and set them off in the woods behind the house or in the empty space of the old landfill in town. It seemed harmless enough to me. When I was an engineer for KISS in my early twenties, we'd fired off enough pyrotechnics for a medium-sized city's July 4 celebration at every show. That always seemed like good fun, and good showmanship. No one got hurt and the crowds loved the spectacle, and most of us were never arrested— at least not on explosives charges! All that happened long before Cubby was born, but it was inevitable that he'd hear about his dad's colorful past. And of course I wasn't the only one in the family with a pyrotechnic background. I may have designed the rocket guitars for KISS, but it was Cubby's mom—Little Bear—who assembled them with me. So he inherited that interest from both sides of the family.

Another thing Cubby had inherited from me was impatience with the public school system. We'd all been frustrated by the way he struggled in school. Major challenges in reading and organization were dismissed by teachers, who called him lazy—the same thing they had said about me. The previous year, after concluding that Amherst Regional High School had nothing to teach him about advanced chemistry, Cubby had dropped out and continued studying on his own.

I'd been relieved when he immediately enrolled at nearby Holyoke Community College after studying hard to get his GED. "I can graduate faster that way," he assured me. The speed with which he now learned math and physics was impressive. And he needed it to understand the graduate-level textbooks he was devouring. By his eighteenth birthday, the stuff he was reading was far beyond anything I had ever studied. That gave me hope that he'd found a path to a future career. I hoped he'd complete college and do better than I ever had.

Cubby didn't have anyone to talk to about his high-level chemistry and physics questions, so he went online. He founded a discussion forum and soon had thousands of threads going on all sorts of topics. He also made some videos of his experiments and uploaded them to YouTube in the fall of 2007. That proved to be his undoing.

A few months later, the Bureau of Alcohol, Tobacco, Firearms and Explosives (ATF) went looking for him at Holyoke Community College. After a short discussion, Cubby led two federal agents and a state trooper to his lab, which by now was in the basement of his mother's house.

The agents were soon joined by fifty of their friends and associates, and they spent the next few days removing the chemicals from my son's lab. On the eve of their arrival, the ATF agent in charge said to me, "Mr. Robison, I want to assure you that the federal government has no criminal interest in your son. We just want to clean up anything dangerous. Somewhere in the United States, every year, we find a Boy Scout genius with a chemistry set, and this is your year."*

Unfortunately, where the federal agents saw a bright kid with some dangerous chemicals, the local prosecutor saw a chance for fame and publicity—her own. She portrayed my son as a budding terrorist and herself as the community's savior. This was despite the fact that no one had complained about Cubby, he'd never been in trouble, and nothing was damaged by his actions.

Two months after the raid she charged him with three counts of malicious explosion—a felony that had formerly been reserved for mobsters who blow one another up with car bombs or criminals who throw grenades through windows when things turn ugly. My teenage son was facing sixty years in prison if convicted.

After the raid, Cubby and I realized that we both underestimated the response to his online activities. I'd known about the videos and had been a little bit concerned that viewers would misinterpret what he'd been doing, but even I failed to anticipate the venomous attack of the local prosecutor, who had never even spoken to my son before filing felony charges.

That showed me just how disconnected we both were from the rest of the world, and I realized that autism probably had a lot to do with that. Perhaps a more aware dad would have better understood the potential problems with Cubby's online postings. I'd always been oblivious to much

---

* The full story was told in my 2012 memoir, *Raising Cubby*.

of what went on around me, which was explained when I got my own Asperger's diagnosis. Now a psychologist had confirmed that Cubby was on the spectrum too.

Autism surely had a lot to do with Cubby's current troubles, but I was adamant that he had really done nothing wrong. Stockpiling explosives for nefarious purposes is clearly illegal, but experimenting with household chemicals in your own backyard should not carry the threat of a prison sentence. Many of history's greatest chemists had started out experimenting just like my son had. The prosecutor had twisted Cubby's actions crazily out of proportion. Clearly one or the other of us was very out of touch with public mores. Only time—and a jury trial—would tell who it was.

Needless to say, the looming court date was a source of stress for all of us. The basement raid happened just as I was signing up for the TMS study, and the trial was not scheduled until the spring of 2009—over a year after the raid. I was not the only one who welcomed the distraction of the TMS study!

If brain stimulation had the potential to awaken things in me, and improve my life, perhaps it might do the same for Cubby. He was fascinated by my experience with music, and he began talking about his own senses. We listened to songs together and decided he might not be perceiving the same level of detail as I was—at least he wasn't able to articulate it. But he was young, and he had never worked in music. Maybe he was hearing even more than I did and just didn't know how to describe it. That's always the problem when two individuals compare perceptions.

Cubby went through the same process of filling out consent forms and signing his life away as I had, and he got the same thorough evaluation. He spent considerable time comparing his brain MRI images to mine, as well as the results of another part of the admissions process—the IQ testing.

The institutional review board had told the science team that everyone they recruited had to have an IQ over 70. They set that threshold because they felt it was important for subjects to be able to describe their experiences, and they thought that was the minimum IQ for that ability to be assured. They gave each volunteer a long-form IQ test that took a psychologist all afternoon to administer. Each of us assembled puzzles, recognized words, and generally demonstrated our brainpower. We definitely got the hundred-dollar treatment there—no quickie Internet IQ testing for this program! At the end Cubby insisted on knowing his score and mine, and ever since, he has proudly proclaimed himself "four points smarter than Dad."

I had always told him he was smarter than a houseplant, and that was finally proven true. It turned out that everyone who volunteered for the study was well above the threshold. With a study population mostly drawn from the Harvard and MIT communities, our group's average IQ was 122. Alvaro and I talked about that and what it might mean for the research. He hoped our pool of bright subjects would be able to articulate what we felt and that the analogy of using zippier computers to get the job done faster would prove apt.

Like me, Cubby was eager to start the actual TMS. He received his stimulations in random order—like everyone in the study—which means his first stimulation wasn't necessarily the same as mine. His first couple of sessions didn't seem to have much effect. We talked, and I watched him closely, but nothing seemed to be different. Then came the third session. After that one, we walked down to the Starbucks, the same one where I'd howled at the ambulance.

As we sat there, something about Cubby seemed different. He was watching things much more intently. After a moment he said so himself. "That's weird! I can see all kinds of detail out there," he announced. When I looked where he was pointing, all I saw was a street full of cars. It didn't look like anything special to me. "Well, it does to me," he insisted. "It's like someone turned up the screen resolution in my eyes. I just went from low-def to HD." After a pause, he said, "I hear more too. I can pick out each of the cars as it drives past."

Some of the car sounds stood out for me too, but I certainly couldn't differentiate all of them. When I said that to my son, he was quick to assure me, "I can." Then he reminded me of those four IQ points.

That sensitivity stayed with him for a while. "There's more color," he said a few days later, and he saw new and interesting patterns everywhere we went. He picked out license plate numbers on passing cars and shapes in brickwork mortar. When I played him some old live recordings he could now tell me how many singers were in the background chorus, and he'd pick up the transitions onstage. I'd never observed that sort of sensitivity in him before.

Both of us found it very interesting that TMS expanded our range of senses, but differently. I got a better sense of hearing, and he got sharper vision. "But not sharper like getting new glasses," he assured me. "Sharper like seeing more things."

Yet he didn't describe any of the more dramatic emotions I had experienced. I wondered if he might have been feeling them but just wasn't saying anything. "I feel a little sharper," he said, and that was pretty much it. When I was young, there had been times when I'd done that, because I couldn't fully articulate what I was feeling. I wondered if Cubby, at eighteen, felt the same.

Seeing into People

THE NIGHT THAT the music came alive for me was a transcendent experience. Even though the energy in my head had dissipated by morning, I remained powerfully affected long after. If I had any doubt about the power of TMS, that one night put it to rest.

The TMS had brought back an ability I thought I'd lost forever—seeing deeply into music—and added a layer of emotional understanding that I'd never known before. The combination left me a nervous wreck—crying at ordinary news stories—but it had also showed me a beauty in art and sound that I'd never known before. The researchers all agreed that my basic ability to "see" music had been inside me all along—I'd used it years before and described it in my writing—and somehow the TMS had set it free. But they didn't know how that had happened or what might come next as a result. It might even have surprised them more than it surprised me.

Alvaro agreed that the music experience was wonderful, but he said it could just as easily have been awful. He had cautioned me about that in the beginning. He'd been very careful to develop his theories about where to aim the TMS for an intended effect, but there was no way to anticipate what direction my thoughts would go in when the energy arrived.

I asked him what he meant by "awful." Would I see monsters and demons? Would I want to hurl myself off the side of a cliff? "I don't think anything that dramatic would have happened," he reassured me. "You felt elation and wonder. But what if you'd felt worry, fear, and anxiety instead?"

The idea that TMS aftereffects might be the stuff of nightmares was very disturbing. His suggestion that any unexpected outcome was just as likely to be bad as good was particularly worrisome because his reasoning made sense to me. The emotions I tended to feel most strongly in life were the bad ones, the things that brought me down. I had often wondered if that could be an evolutionary trait. Failing to sense a good thing might keep us from a momentary joy, but failing to sense a bad thing could get us killed.

Could there be a mechanism in my brain that tilted the odds in my favor? That had occurred to me earlier, in the context of my career. Why had I been successful while other autistics who tested better than I did struggled to find jobs?

To my amazement, Alvaro did not dismiss that as a crazy notion. His answer surprised me.

"You've done a lot of things in your life. Most of the time, you educated yourself and made your own way. And you succeeded. Most of us are happy to succeed at one thing. But you succeeded in music, in electronic games, in car mechanics, in photography, and now in book writing. If you were just relying on luck, half or even all those things would be failures. You thought you failed at everything at the time, but you hadn't. You didn't succeed randomly. You succeeded because the engineering, the technology, the writing were . . . correct. You did it right, without any formal training. Why did things work out for you so many times? Who knows? Maybe there is something in your mind that guides you to choices that will work. We just don't know."

Dr. Nancy Minshew said something similar when I spoke to her some months later at an autism science conference. "Some people just make the right choice consistently. It's not luck, and no one knows how to define it. We should study the brains of successful autistics to see if we can find the answer." The notion that I was successful was flattering, though I didn't see myself consciously following any kind of invisible optimal path. All I did was complete the work at hand as best I could. Was there something to be learned from that?

Many people do the same thing, but what Alvaro took notice of was that I had done that successfully in multiple areas of endeavor—something I had not really considered before. I wondered what the common thread was that allowed me to succeed. Maybe I should have become obsessed with mastering professional blackjack and poker when I was younger. I guessed that so-called success trait was what Nancy wanted to find.

A few people have suggested that maybe I was just drawn to what I was good at, the areas where I'd have the best chances of success, but I don't see it that way. How could I have anticipated creating those KISS guitars the first time I picked up a transistor and gazed at it in wonder? Alvaro was right—the odds against my sequential successes being simple good luck were very long.

Even with those encouraging thoughts, the notion that TMS might cause me to spin off in a negative direction had me worried. I'd already drawn a mental analogy between my musical hallucinations and a drug experience. I well remembered the stories of musician friends who took psychedelics and ended up on bad trips—nights of horror that left some of them scarred for years. But I didn't say anything, because I was ashamed and embarrassed by those thoughts. I had to wear a brave face for the re-searchers, and that's what I did when I arrived at the lab for our next stimulation. And I'd nearly forgotten that a TV crew would be there, to film some of the day's session.

Several weeks earlier, a television producer had contacted me on behalf of Dr. Norman Doidge, whose book I'd read before first meeting Lindsay. Doidge is a Canadian psychiatrist who studies the powers of the mind, with particular interest in neuroplasticity, and he was filming a TV series to complement his book. He wanted Alvaro and me to be part of it—to go on camera and talk about brain stimulation. Medical research is usually conducted in private, but my situation was already unconventional—I had met Lindsay in a public venue, and I was already speaking out online about my interest in her research. I had chosen to give up some of my own privacy, in the hope that others might see the promise of what I believed could be a life-changing therapy.

That was why I'd said yes when the film crew had reached out to me. My friend Michael Wilcox had also agreed to be involved, and the crew planned to shoot footage of both of us receiving TMS that day. Having the Canadian Broadcasting Corporation (CBC) television crew in the lab in the midst of our experiments complicated things, but I really enjoyed meeting and talking to Dr. Doidge. I'd found his writing fascinating and inspiring, and we spoke at some length about his ideas throughout the day. He described techniques for changing our brains, and people who had done so in amazing ways. The CBC crew filmed both Michael and me as we underwent our stimulation sessions, and shooting didn't finish till after seven P.M. Martha and Cubby were with me, and by the time we were done, they were hungry. And of course I'm always hungry. Lindsay, Shirley, and Erica from the lab decided to join us for pizza down the street.

Alvaro went home, and Dr. Doidge and the film crew had other plans. That was just as well, given the strangeness that was about to unfold.

The six of us who remained began talking once we were seated at the restaurant. I'm generally a pretty circumspect person, but I'm no saint. During my rock and roll days I experimented with drugs and drinking, like most everyone else in that universe. However, I'd gotten all that out of my system decades ago. Today, conversation with me rarely gets edgier than car engines and off-roading. I've always been insecure about how others see me, so I'm generally cautious about what I say. For some reason, everything was different that night. Whatever inhibition I normally have was gone, and I began recalling hallucinogenic experiences from the seventies, things I hadn't thought of in years. As everyone else munched their pizza in astonishment, I regaled the group with a story about eating a handful of mushrooms during a James Montgomery Blues Band show at the Shaboo Inn in Willimantic, Connecticut, and then feeling an overpowering urge to drive my friend's car to Canada because I needed desperately to pee on Canadian soil. I couldn't pee till I got to Canada, no exceptions. We were facing a genuine crisis. Several friends—also high on mushrooms—had driven north with me, and they must have shared my delusion, because we crossed the border at Rock Island, Quebec, drove a mile into the country, and got out and peed in a line by the roadside.

A story about eating psychedelic mushrooms and then driving 275 miles to take a piss is one that any sensible adult would be embarrassed to relate, especially in front of a group of esteemed scientists and an impressionable eighteen-year-old son. I'd never said a peep about it before that night; in fact, the incident had faded in my mind like an old tie-dyed T-shirt. But somehow it popped into my head again and I couldn't seem to stop myself from telling them the tale in all its glory in that pizza parlor.

That wasn't even the end of the story. Once we were done peeing, we'd picked up some rocks as souvenirs of Canada and, mission accomplished, gotten back into our VW Squareback, turned around, and started for home. We crossed the border, and it was smooth sailing for about five minutes. Then four Border Patrol cars roared out of the dark, pulled us over, and the lawmen arrested us for driving back through the border at four in the morning without checking in. We had been so sure that the border station was closed!

The cops brought us back to the station, searched us, and searched the car. They didn't find much more than the Canadian rocks, and they let us keep them, which was a comfort. At first they thought we were up to something nefarious, but they soon figured we were just crazy kids and they let us go home. Someone trying the same thing today probably wouldn't be so lucky.

Cubby thought the story was funny, and Martha whispered that it was a strange thing for me to be talking about, especially with people I didn't know well. But howling along with the ambulance had been stranger, so she just sat there waiting to see what came out of my mouth next. Maybe everyone feels the urge to pee on foreign soil at some point. But if the scientists felt that way, no one admitted it.

Another night's memory came back to me, and I related it in vivid detail too. That time, I ate mushrooms while lounging by the woodstove at a friend's house in North Amherst near where I lived. The psilocybin took hold, and the next thing I knew I was flying fast above a cold, high western desert. The sky was a beautiful purple, filled with stars and waves of light, like the aurora borealis but filmier and infinitely more beautiful. As a creature of the sky, I flew for a long time, soaking up energy from the colors of the darkness and the rippling of the light. Far below me, the desert was dark and empty. Swooping gracefully toward the earth I found myself jolted back into the real word with a bang. There I was on my old Honda 750, in real life, riding through the gap in the sawhorse barrier that marked the endpoint of the (still being built) Interstate 91 near St. Johnsbury, Vermont. I barely stopped my motorcycle before plowing into the rough gravel after the pavement came to an end in a few hundred feet. How I avoided crashing between Amherst and there, I'll never know.

St. Johnsbury is a good three-hour ride from where I'd started out in Massachusetts earlier in the evening. I have no recollection of getting from one place to the other; all I know is that I miraculously made it safely through the deserts of New Mexico and the highways of Vermont all in one night. Once the bike skidded to a halt, I took a deep breath, killed the engine, and looked up into the night sky. Northern Vermont hasn't got any cities to fill the darkness with light. There's no pollution to color the air, and that night there were no clouds. Nothing but stars, airplanes, dragons, and the other creatures of the dark. Millions of pinpoints twinkled back at me, but I never saw the beautiful purple and the waves of light again. It was a long cold ride back home, and I didn't make it there until morning.

As I told these stories I realized I must be giving them the impression that I'd had a wild and drug-crazed youth. That wasn't true at all, and I made a point of saying so, for whatever good it did. I'd only done mushrooms a handful of times, and never anything stronger. Those few mushroom experiences were bizarre enough to put me off stuff like that for good, even though I'd never had a bad trip. Most people I knew in those days smoked a little dope and drank some beer, but the aftermath of both usually left me anxious or embarrassed. Relating to the world was hard enough when I was sober; doing so high or drunk was more than I could handle.

It seemed awfully strange that those memories would come flooding back just after the TMS session. The images came out of nowhere and shouldered their way to the front of my consciousness. These were not seminal moments in my life; they were experiences that I'd nearly forgotten. They weren't cherished memories, like those of my experience with music. All I could figure was that TMS had brought long-buried memories to the forefront of my mind, but why that might have happened remained a mystery, to me and to the scientists.

It was late when we got home, and I didn't get to bed until 12:30. I lay down next to Martha and closed my eyes. That was when the strangeness hit me.

The world started moving.

At first, I didn't connect the movement with TMS. Frankly, my first thought was that I was drunk, because that's what it felt like. However, I'd only had iced tea at dinner. Then I got scared. *Am I having a stroke?* The world was swirling and twisting, slowly but steadily. With a start, I opened my eyes. Just like that, the motion stopped. When I closed my eyes, it began again.

Anyone who's been falling-down drunk would recognize the feeling. The spins. It's what happens five minutes before you throw up. This time, however, there was no nausea. And the long and eventful night was just beginning.

As the world spun vertiginously around me, images began playing in my mind. Vignettes of the day just passed were followed by crystal-clear recollections of my early childhood. Suddenly, I was two years old, sitting underneath the seat of the white rocking chair on my grandmother's front porch on her farm in Georgia. I could hear her saying, "Watch out, John Elder! Keep those little fingers away from the rockers!" I gazed out across the porch, and then I was back in Alvaro's lab earlier that day, looking at the doctors. It was like I was a character in a movie of my life, but with no chronological sequence, no beginning and no end.

Next I found myself in the hall at Beth Israel with Dr. Doidge, who was looking at me intently and saying, "Can't you see the energy around certain people? Whatever you may think of Obama, can't you see his charisma?" The TMS experiments were unfolding during the run-up to the 2008 elections, and Obama was all over the news. And in the dream I thought, *He's right, I do see something. How do I explain it to him?* Then I opened my eyes, and I was in bed in Amherst.

I saw that Martha was sound asleep at my side. Whatever conversation Dr. Doidge and I had been having had not woken her up. Anxious as she was about the whole TMS process, I didn't want to startle her with this in the middle of the night. I resolved to keep quiet and tried to remember exactly what we'd talked about that day. It seemed likely that we did have a conversation about Obama. Or maybe I imagined the whole thing. I do know this: the Dr. Doidge I was conversing with in my head was one heck of a personality. He was like a real-life Yoda. My hallucinations that night were so vivid that they've left me unsure about what really transpired between us.

As I lay there, a fresh thought jolted my mind. *Where are the inflatable life jackets for our boat?* I could not remember. Wherever they were, I knew I needed to change the inflation cartridges. A little voice was telling me that the major cause of failure of inflatable life vests is that the cartridges expire and they don't work when they are needed.

Is the house sinking? Sometimes animals sense stuff like this in advance. *A life jacket can't save you if you're not wearing it! I'd better go put one on.*

Opening my eyes again, I looked around and marveled at the strange stuff that was happening. The rational side of my mind was struggling to surface and was rejecting the possibility that the house was about to founder.

Then I was back in Alvaro's office, talking to Lindsay and Alvaro about diagnoses. "There are tons of adults with no diagnosis," I said, "and they're going to be reluctant to get one, because they'll be branded by the insurance companies. Marked as autistic." I couldn't tell if I was talking in my sleep, imagining this conversation in my mind, or recalling something from earlier in the day. I still don't know.

Then I blinked, literally, as if I'd been startled, and I was in the TMS lab with the whole crew. In my dream, I looked at the clock and saw it was 6:00 P.M. The extra time spent filming had made us late, and the scientists were worried. The stimulations they were filming were part of a real study, not a sham for television, Lindsay warned them. "You guys have to move out of the way, because the effects of this TMS will fade fifteen minutes after stimulation and we have to test quickly." I'd heard that enough times to know it wasn't my imagination, but then why was all of this coming back to me at three in the morning?

A really strange idea hit me just then. *Tonight, it's all about connections. That's what it is. Connections. An idea pops into my head, and then I make an association, and another, and another. And the floor tilts and I slide into the next idea. It's as if there's a mass of spaghetti in my brain and it's sprouting tendrils, weaving itself together. The end result? I have no idea.*

That was how the night passed. Even though nothing bad happened in my waking-hallucinating-dreaming condition, the effect was disturbing, and I sat up in a state of high anxiety at 4:05 in the morning. As I came to my senses, the feeling of anxiety vanished, but I remained alert. Unless I had dreamed the anxiety too? Dream and reality were pretty mixed up that night, and this wouldn't be the first time I woke up anxious over something in a dream.

Walking across a dark and quiet house, I went upstairs to my study, where I have a commanding view of the driveway and the woods around us. It was still dark outside; I couldn't see a thing. I sat there quietly, waiting to see what would happen. Nothing did. When I looked at the clock it read 5:17. Just like that, an hour and twelve minutes had vanished. It was pitch-black, but a few birds had started chirping. Or were they a hallucination? *Shit,* I thought to myself, *I better find out. If I'm still hallucinating, the scientists will want to know.*

When I opened the window I could feel the cool of the darkness before the dawn, and the sounds and smells of woodland spring came pouring in. With a sigh of relief I realized that the birdsongs I was hearing were real. Closing the window, I had a new and unsettling thought.

Sometimes, when the birds started singing, it was because a dangerous animal had entered the neighborhood. *Should I get the shotgun?* I wondered. But I was safe in the house, up on the third floor, and I hadn't heard of any animal home invasions in Amherst.

What I felt was a brief chill, a whiff of danger, but it passed in a moment. All my life I had been anxious, and I wondered why I was not in a state of total panic now. *How can this be happening, with me staying calm?* The night had been one strange experience after another, yet I was resolutely upbeat. *What a remarkable development,* I thought. Perhaps that insight released me, because I opened the window again, sniffed the darkness one more time, and padded downstairs to bed.

Once asleep I was out like a log. There were no more hallucinations, and I can't remember dreaming. When I finally awoke—at noon—I opened my eyes to a gentle motion, as if I were on a boat, rocking at anchor in a harbor. *What is wrong with me?* The world around me looked the same, but the ripples I was feeling would not stop. Every time I closed my eyes, things moved faster.

I got up, got dressed, and driving very carefully I made it to work without incident. The guys were all in the shop as usual, and our service manager Maribeth sat behind the counter, talking to customers. Everywhere I looked was bustling with activity, which I took in as I looked around.

Then I made eye contact with Eddie, who worked in the shop, and a thought struck me with overwhelming intensity: *He has the most beautiful brown eyes. I wonder why I never noticed that before?* A moment later, a voice inside my head said, *What's going on? You've never noticed anyone's eyes before! What's happened to you?*

Indeed, I had turned away from him, shocked by the power of the feeling and overcome by a mix of emotions I could not even name. Before that moment, I'd been uncomfortable even looking in someone's eyes. Now that had changed. Avoiding eye contact was such an ingrained part of my life that it gave me the name for my first book. Staggered, I turned my attention to one of our customers, who'd been waiting to tell me about her broken car.

As she spoke, her face began to tell its own story. I wasn't even hearing her words, but her feelings shone through clearly. As she calmly described the symptoms of her car's misbehavior, her eyes were saying, *I'm really worried about what this will cost because I'm insecure about my job and whether I can afford to fix the car, but I need it to get to work so what's going to happen?*

I responded immediately, with reassurance. "Don't worry," I told her, "the problem you are describing sounds like a pretty small thing to fix." She immediately relaxed. The whole thing happened so quickly and naturally that it took me a moment to realize the significance of what had transpired.

Somehow I had read the expressions in her face and answered them instinctively—and correctly. Most people take such abilities for granted, but I had a lifetime of experience missing those cues and saying the wrong things—sometimes the worst possible things—in response to the logical words others spoke to me. A few days before, I'd have listened to her story and said, "Huunh! Have to bring the car in and see what's the matter." Her fear and anxiety would not have made any impression on me at all; consequently, it would not have occurred to me to reassure her.

You are seeing into their souls, the little voice said. Just then, I felt another flood of emotion and I had to step outside as the wave washed over me. Human eyes had become windows, and the surge of emotion pouring from them was very powerful. And somehow my new ability felt completely instinctive and natural, as if it had been there forever.

How will I make it through the day? I wondered. As the shop grew busier and workers and customers surrounded me, the emotions I was sensing went from a trickle to an overwhelming torrent. Noticing Eddie's eyes for the first time was remarkable, but experiencing the collective emotional energy of a small crowd and feeling each person's hope, fear, excitement, and worry was just as disabling as being blind to it. The nature of my disability was effectively turned upside down. Originally I'd been oblivious and my emotional world was comparatively silent. Now it was full sensory overload with a cacophony of distinct emotions swelling like a discordant symphony all around me.

When I could focus on one person it was as if I had emotional ESP, but thinking back I know it must have been the newly awakened, fuller power of my mind. The emotional onslaught was unfamiliar and overwhelming. And the day was just beginning.

By some strange twist of fate, I'd been booked as the keynote speaker for the Massachusetts Medical Society's annual meeting that night. That was another of the invitations that came my way after the release of *Look Me in the Eye*. Speaking to three hundred seasoned middle-aged doctors, I felt as if I were looking into their very souls, where I saw their fears and hopes and dreams. People who were in the audience later told me they'd never experienced anything like the connection I made with them that night. Somehow, I touched them with my words in a way I had never known I was capable of doing.

When I spoke I told them about being autistic, and alone. Then I talked about TMS, my hopes for change, and my dreams for the future. My exact words are lost to memory, but there is one thing I will never forget: I spoke from the heart to a group of hardened medical practitioners—doctors who had seen and heard it all—and I made them cry. I saw the tears and felt their empathy reaching back out to me. My doctor friends in the audience still talk about that night.

Dave had been seated at my table for the dinner that preceded my talk. "The fellow sitting next to me was that older doctor who'd gotten an award from the medical society for fifty years of service. And he was listening to you with tears running down his face. It was remarkable," he told me.

By the time the event was over I was worn out—I felt as if I'd hiked ten miles. When I got home and told my wife what had happened, she said, "Well, you won't need me anymore." I staggered as if I'd been hit. As much as I loved her, and the life we'd built, I knew she was right, that something fundamental had changed in me, and as a result, it had changed in us. There was nothing I could say in response, but I felt unutterably sad.

Hallucinations
and Reality

THE NEUROLOGIST Oliver Sacks wrote a book called *Hallucinations,* in which he described being partly blind—according to tests—yet having his field of vision filled in by his imagination. He described a similar phenomenon with his patients, with respect to sounds—hearing things that were not there, or hearing things different from reality.

Some of his patients heard trains when no train was present. Others heard the voices of departed loved ones. In his book, Dr. Sacks attempts to account for those hallucinations medically. Others might explain them spiritually. Dr. Sacks makes the point very eloquently that anyone is capable of imagining things that are not there under the right circumstances, and such visions may be visual, auditory, or even olfactory. The mind apparently has a great power of creation. One thing Dr. Sacks does not address is the difference between outright imagined fantasy and vivid and sudden recollection. Try as I might, I could never figure out whether my musical hallucinations were imagination, memory, or a mix of both.

My experience of seeing into people—looking in their eyes and sensing their thoughts—was completely real, as far as I could tell. It didn't seem like hallucination; I came to believe it was more a dramatic expansion of a sense I had always had, though at a very low level.

Dr. Sacks avoids any territory that cannot be explained by science, yet there are countless stories that science has no answers for. I'd had such an experience myself, ten years earlier. I was driving on Route 128, out by Boston, when I had the feeling that something terrible had happened to my father. When I dialed his cellphone he didn't answer. That gave me a queasy feeling, but I wasn't sure what to do, so I continued on my way. Forty-five anxious minutes later, my phone rang. It was the state police. "Your parents have been in a crash," the dispatcher told me. "They're still alive, but you'd better go to the hospital now." I wheeled my car around and stepped on the gas, realizing that I could have followed my original instinct and been halfway to the hospital by that time. I never questioned premonitions after that night, though I've no idea how to explain what I had experienced.

Shamans describe hallucinations as a vehicle for leaving the body and traveling to distant places; they see them as a means of making connections with things we can't reach otherwise. They're lessons, or learning experiences. Is that what mine were? Were they also a way for my mind to reach inside itself and make new connections? Maybe the shamans were onto something all along that traditional medicine is only now on the cusp of discovering.

As you might imagine, I did a lot of thinking about the powerful experiences of that night and day. That afternoon I had received TMS, and for the next little while, nothing obvious had happened. Four hours later I began thinking of—but not experiencing—hallucinations, and talked about them all through dinner. Then I went home, and starting just after midnight, I had visions almost until dawn. Michael had gone home right after his session, and I wondered what had happened to him, but we had agreed not to discuss it till the study ended. When we did talk, a few months later, I was disappointed to hear that he'd felt a sharpening of his senses, but the hallucinations that hit me so forcefully didn't come to him at all. Was that because it took more to change him? I asked Alvaro, but he didn't know. Also, Alvaro reminded me that they didn't stimulate the same areas in each of us that day because the order of the target areas was randomized.

It was as if the TMS had spoken to me by putting my mind on the track of hallucinating and then making it happen. That was very strange. The events of that night had hit me with enough force that before dawn I had written and sent a thousand-word missive to Alvaro as I'd sat upstairs in the dark in my study. He was startled to find it in his inbox when he woke up the next morning.

In my email and afterward, I tried to make sense of it all. Had the scientists just happened to hit a "right spot" in my brain, releasing a flood of imagination and memory? Was there a "secret switch" for that behind my right temple? Or were the hallucinations the cumulative result of energy and changes over the entire series of TMS sessions? Whatever the trigger, the aftermath of the experience was shaping up to be more noticeable with every passing day.

The results of every TMS session seemed to follow a similar pattern. There was an immediate effect, which the scientists tried to capture with their testing. Then there was a short-term effect that I felt in the hours after. Finally, the longer-term effects arrived, and some remain with me now as I write this story, years later. Will they last forever? Only time will tell.

In my five A.M. email to Alvaro, I'd described my vision of the mass of spaghetti in my brain that was sprouting tendrils and weaving itself together. As odd and science-fiction-like as that image was, it may be the best way of articulating exactly what was going on in my head all through that night. Alvaro told me that our brains are making new connections all the time, but that the rewiring is subtle. However, the changes that I'd experienced from TMS and its aftereffects were much more profound and far-reaching than the minor rewiring the scientists believed the brain could do overnight. But perhaps they underestimated the brain's ability to reconfigure itself, or maybe our minds have many alternate pathways and the TMS brought some of them into use.

The ability to hallucinate is latent in all of us. When we take certain drugs like I did the night of my trip to Canada, chemicals get into the synapses between our neurons and trigger strange visions. People with psychosis also hallucinate, without any drugs at all. That happened to my mother—she saw demons and monsters when I was a kid. Her doctor called that "having a psychotic break," but that clinical term does not even come close to describing how it felt to me, at age thirteen. It was terrifying. Those visions of my mother have remained with me all my life, and they probably deterred me from more enthusiastic experimentation with drugs. But the hallucinations I'd experienced weren't because I had gone crazy, and no chemicals had been added to my brain. The only addition was energy.

Is energy alone enough to precipitate change? When I'd asked him, Alvaro certainly thought it was. "Think about our depression treatment," he told me. "We are replacing psychiatric medications with TMS energy and getting excellent results." That raises a new and vital question. If pure energy can change—and possibly "fix"—the mind, can the mind then direct the body to fix itself? That seems like a valid line of questioning, one that offers the promise that one day, medically validated energy-based therapy may supplant drugs for certain treatments.

In response to my questions, Alvaro shared with me his dream of creating a brain health center at Beth Israel Hospital. Many of us exercise our muscles, but almost no one exercises her brain. Yet brain health is more important than the health of any of our muscles, except maybe the heart. Diet, activity, and many other factors affect brain health. "I believe a healthy brain will maintain a healthier body," Alvaro assured me.

"A healthier brain would not necessarily prevent us from getting cancer," he told me, "but it would help us to adapt better as we age, and that versatility will allow us to make the best of the bodies we have, for a longer period of time."

We talked about what role TMS might play in brain health, and he described a future in which brain stimulation might guide a person's brain down a more optimal path as his body aged. That sounded like some pretty high-level stuff, but he assured me it was possible. His words made me suspect that something similar was happening to me now. "Your brain is always optimizing itself," Alvaro told me, "and we hope TMS guides it to do that job better."

The delayed effect that I experienced really adds to the mystery. When we swallow drugs, we know some time is required for the chemicals to move into the bloodstream and from there to the synapses in the brain. So a delay of minutes to hours is not surprising when we take medication, as everyone who's taken a pill has experienced. There's also the issue of dosage with medication. You may need to take several doses to build up the chemical levels in your body to the point at which something happens. That might take a few days or even a week. And different people respond differently to doses of medication, just as they do with alcohol or drugs. One man's gentle buzz is another person's falling down drunk.

I observed that with the antidepressants Martha had taken. When she switched to a different drug, the pills wouldn't seem to do anything at all for a week, sometimes longer, and then they would start working. Her psychiatrist would always warn her to expect that delay. When the drugs started working—if they did at all—the effects might be mild or dramatic. Waiting to see which would happen was always discomfiting.

There were times when Martha's depression kept her from getting out of bed for days, and we wondered if TMS might help her depression, but the depression research trials had ended at Beth Israel and regular patient treatments hadn't yet begun. Besides, she was somewhat frightened by the effects she had seen in me. Energy seemed like pretty powerful stuff— especially compared to medication that didn't always seem to work—but perhaps it was too powerful.

Unlike drugs, TMS delivers energy directly to the brain, so we expected that its effects would be immediate. That was why the researchers wanted me to move quickly to do the after testing. But it didn't really work out that way. Though there were days in the lab when I felt like I'd breezed through their tests after TMS, and other days when I struggled with every question, the immediate effects the scientists were trying to capture with their tests paled in comparison to the bigger changes that unfolded later. In some ways, a drug trip seemed to be an apt comparison. You swallow the drugs, and nothing happens. You walk around, and still nothing happens. You turn to the guy next to you and say, "Hey, man, this stuff is no good. Nothing's happening." Then all of a sudden you are on the floor, with long-fanged creatures swarming at you from the ceiling.

Alvaro said he'd seen a cumulative effect using TMS in other research, but the mechanism by which it happened was not fully understood. Later, Lindsay would tell me about homeostatic mechanisms that she called "emergency brakes inside your brain." They keep our brains from changing too much too fast and becoming unstable. "Because of that," she explained, "you need to slowly but surely shift the synapses in your brain to be more plastic in the direction that you are aiming for. These mechanisms won't allow you to do it all at once to ensure that your brain has some stability."

I had initially imagined the TMS energy as recharging a chemical battery in my head, but it's not really that simple. The TMS changes one circuit in the brain, and that alters two more. Those two affect ten others, and those change five hundred more. The interconnected effects are so complex that we can't begin to fully map them with the level of technology available today. That process of sequential change must take some time and it's a possible explanation for some of the delayed effects I experienced.

Everyone has seen how static electricity makes our hair stand on end, or makes paper stick to our sweaters. That effect is coming from excess energy that we pick up walking around. Something similar may happen, on a submicroscopic scale, inside our heads. When the TMS coil fires its pulses, the energy induced on the wires between our neurons may grab extra ions and pull them onto the biological wires between neurons. When that happens, other neurons may fire unexpectedly, and strange things like hallucinations may occur and new and unexpected reactions may be initiated. Current studies support this idea, and Alvaro, Lindsay, and I had quite a few roundtable discussions about it, but it will be years before such hypotheses can be tested and evaluated. When I recalled Alvaro's broad hopes and objectives, it seemed as if he envisioned TMS as heightening my senses by suppressing the suppression. That is what I hoped was happening, as opposed to something bad or crazy. There were not yet any published studies of TMS and autism, but there were a number of professional journal articles on TMS and depression, and I scanned those for ideas that might be relevant to my own situation.

One topic I ran across was that of "sham" TMS, where research subjects were made to think they were getting TMS when in fact they were not. Sham TMS was often used as a kind of control, in an effort to determine whether a volunteer's response to TMS was real. One depression researcher had written, "Most participants reported positive effects, but the ratio of patients reporting positive effects from 'true' TMS as opposed to 'sham' TMS was about the same. More study is needed." That was the same thing researchers sometimes found in drug trials, where volunteers who got the placebo drug often reported positive effects. Essentially, those who got pretend TMS reported almost as much benefit from that treatment as the subjects who received the real intervention. I wondered what that might mean, and I realized there were several possible answers.

First of all, there's a large body of evidence to suggest that belief in a cure makes a cure more likely, and more successful. Pharmaceutical researchers grapple with this issue all the time, as the success rate of placebos and real pills is often surprisingly similar. A few years ago researchers dismissed the placebo response as imaginary and of no medical value. Yet recent studies have shown the opposite: not only can the placebo effect deliver real lasting benefits, but the belief that something is making us sicker can actually lead to real deterioration, even death.

Saying "It's all in your mind" used to be a way to dismiss unexpected effects, but all of psychiatry is in the mind, so that dismissal is not so relevant when it comes to how we feel. It's possible that some of us went into the TMS study believing in its power while others were largely neutral and we were now having different experiences as a result. Michael thought that was so. My son wasn't sure. He went into the study with a mix of indifference and curiosity. He noticed heightened color and sound sensitivity, but he didn't try to make anything of it as I had. He didn't concentrate on pictures or music, and eventually the changes faded to invisibility. I couldn't talk to the other participants to learn how they felt. Might something similar have happened in the depression studies I read about? When treating a disorder of the mind, "believing it's so" might well be tantamount to "making it so," in a way that's dramatically different from treating chronic disease elsewhere in the body.

There was also another important difference when comparing the placebo effect in TMS and medications. In drug therapy, a placebo pill is truly inert. It's flour, or sugar, something we know does not have curative properties. That's not the case with sham TMS. Scientists create sham TMS in many ways, including by firing the TMS coil into space beside your head, firing it at a different area of your head, or firing it at a low power level.

The thing is, all those sham actions still have the potential for neurological effect. It's more accurate to call them "different" TMS because many of the so-called sham techniques are still delivering stimulation to the brain, albeit to a different area or in different form.

The only way to do an effective sham TMS would be to have the TMS machine make a subdued noise and not deliver any energy at all. But the problem with that is that anyone who has experienced real TMS would immediately know something was wrong. TMS has an unmistakable feel—your facial muscles twitch in a certain way, and you feel the zap as energy hits your head. And I felt the meditative state—the trance—almost every time I got TMS in this study.

My own experience with sham TMS during the study left me puzzled. I didn't know what it was when I experienced it, and I wasn't shouting "It's fake!" but I knew something was off and asked the researchers what had changed to make that particular session different from the others. Here's what I wrote in an email to the scientists after receiving sham TMS in one of the later sessions of the 2008 study:

> My first four TMS sessions were all eventful, each in different ways. Therefore, you may be surprised to hear that I've virtually nothing to report after the May 6 session.
>
> In most of the prior sessions, I felt myself falling into an almost meditative trance. I'd sit there, with an essentially

empty mind, for 30 minutes. Even though my face was twitching, I'd sit there, contented and tension free, and devoid of any internal dialogue.

Yesterday, I did not fall into that state. I remained alert, just as when I arrived. Even though I heard the noise, and felt my face twitch, nothing seemed to happen inside. I know what you're thinking . . . there are no nerves in there to "feel." I understand that, but somehow, I did "feel" other TMS sessions and the more I do it the more I pick it up. The expected feeling in my head just was not there, for whatever spot you hit yesterday.

On the way home, I did feel a slight headache, and I felt the tiniest bit of disorientation, but there was nothing in terms of either perceptible main effect or perceptible side effects.

If this was supposed to be another prime area, I'm sorry to disappoint.

No one told me it was sham TMS until later. All I knew was that it wasn't like the other sessions. If I could recognize the difference—even without knowing why—it highlights a problem for researchers. Pills are pills; you can replace the real ingredients, but the experience of swallowing them is the same. TMS doesn't lend itself to faking.

I certainly never felt like I was faking in any of my responses to TMS. But as dramatic as some of my experiences felt, I did wonder if I was going crazy and imagining things that were not really there. That was particularly true after the experience of feeling as if I could see into the souls of other people as I did at the Massachusetts Medical Society dinner, which came right on the heels of seeing into music and developing my newfound emotional sensitivity. Those were some of the most intense experiences of my life, and I had to wonder whether they meant I was acquiring superpowers, going crazy, or both.

There was precedent for that in my family. My mother had been absolutely certain of the demons and monsters that chased her to Northampton State Hospital thirty years before. Her younger brother—my uncle

Mercer—had been discharged from the service and institutionalized after seeing things that weren't there and having a psychotic break on a navy ship off the coast of Vietnam. The possibility that my experiences might be equally imaginary was frankly terrifying, especially since I could see that it ran in the family, with Mercer and my mom. None of the scientists ever suggested such a thing, but I still could not help wondering.

Validation of the reality of my experience came out of the blue with a mid-study phone call from Shirley. "I'm hoping you can help us," she told me. "We're not supposed to have participants in a study meet each other, but something's happened. After you told me about the results of your most recent stimulations, and how you saw emotions and thoughts in the eyes of people around you, I waited to see if anyone else would describe a similar experience. We stimulated several of our other volunteers in the same area as you, and they didn't report anything. Then today I got a call from a participant who was here for that stimulation yesterday and was beside herself with distress today. She described a lot of the feelings you told me about, but she wasn't excited about them like you were. Instead she was quite upset. I wonder if you could talk to her, because you've had a similar experience but you felt yours was positive, and you might be able to give her some kind of comfort. She's given me permission to connect the two of you." With some trepidation, I agreed to give Shirley's subject a call. "Great," she told me. "Her name is Kim Davies,* and I'll get you the contact info."

I felt a little funny about that, because I assumed all the TMS study subjects knew who I was, yet I knew nothing of most of them. That was the result of me talking TMS on my blog and at lectures, as the public face of the program, while the other participants were hidden behind a wall of medical privacy. The only participants I knew by name were the ones who approached me before or after they joined. And I tried to keep those contacts to a minimum so as not to influence anyone else's experiences or expectations.

Shirley must have been very worried, otherwise she would never have

---

* Name changed to protect her identity.

allowed us lab rats to speak to one another about our TMS experiences, and I worried about what that might portend. I called Kim, and we agreed to rendezvous that evening at a café. I set out a few minutes later for our appointment. Though Kim and I had never met before, we picked each other out right away, and I could tell she was highly anxious. So was I, because I'd considered my TMS experience to be remarkable, but according to Shirley, Kim viewed hers as awful. All the way to the meeting I wondered if her TMS had unleashed some demon of the mind, and if so, whether something similar might soon be in store for me.

When we'd spoken on the phone, Kim told me she was a physician, about my age, from the other end of the state. When I heard she was a doctor I wondered if she knew something I didn't. But when I met her she turned out to be an orthopedist, specializing in bones, not brains, and she didn't understand any more about the workings of the mind than I did. Kim said she'd learned about her own Asperger's after her nephew was diagnosed with autism. Then she read my book and my writings online. "As I was reading, a lot of things you wrote about felt familiar to me," she said. "Then I learned about the Beth Israel TMS study from the Asperger's Association in Boston, where I'd gone to find out more about Asperger's in me." She'd signed up for the study with the hope that she could learn more about herself.

It felt a bit funny talking to her, because she knew a lot about me from my book, whereas I knew nothing about her. All I could do was pay close attention. And I wondered: How many of the other volunteers had been inspired by my example? Flattering as that notion seemed, it was tempered by a real component of worry. *Should I feel responsible for my readers if they've followed me down this road?* That question alone was evidence of a building change within me; I'd never have asked myself such a thing in the past, but I wondered about other people more and more now.

Before I began the study, I'd written a blog post I called "Standing on the Brink," expressing my thoughts on the hope and promise of TMS. In it, I wrote, "I've had people ask me, 'If you could take a pill and get rid of your Asperger's, would you do it?' I've always said, 'No! I'm proud to be Aspergian and I wouldn't change it for anything.'" But I acknowledged that my challenges are as real as my gifts, as much as I wish otherwise. I recognized that some people take medication for anxiety, depression, and the other conditions that often accompany autism. I've never felt they were right for me. "But I've always wondered," I continued, "what if there was an alternative to pills? If I could pick a part of my mind and change it and make it better, would I do it?"

I reread that blog post as I waited to meet Kim. I still believed everything I had written about the hope that TMS offered. If anything, I believed in TMS more, not less, than when I began. But I was also getting a glimpse of the incredible difficulty neurologists would face in deciding how to use the tool.

Since Kim and I hadn't met before that night, I had no idea what she looked like, or where she would be, but we found each other right away. When I speculated on whether that was autistic radar in action, she reminded me there was a photo on my book jacket, and I quieted down in embarrassment.

We discussed her experiences with the stimulations and the exercises before and after. Both of us had received our stimulations in random order, and the ones she'd done first hadn't had much impact. But the most recent one was a beauty. As she described it: "Following the stimulation, I looked at people and was able to accurately read their expressions—something I'd had trouble with before. I heard someone talking and I knew they were being sarcastic; I looked at people and I could read how they were feeling."

My emotions had been running high ever since that first night I saw the music, and hearing Kim validate my own experience was almost too much for me to bear. I had to close my eyes and compose myself.

The same thing had happened to me!

She thought the direct effects had faded, just as they had for me. But she was still reeling from the aftershock. Like me, she had gotten a glimpse of the emotional world in full color, and after that, life is never the same. But for some reason, our interpretations were—at least initially—virtually opposite. Which of us was right? Did it have to be one or the other?

Australian philosopher Frank Jackson devised a much debated thought experiment that seems to describe our experiences. It's called the Mary's Room analogy.* Lindsay had told me about it one afternoon at the lab. In it, Mary is a scientist specializing in how the brain processes visual data who is forced to investigate the world from a black-and-white room via a monochrome television monitor. She acquires all the technical information there is to obtain about the real world outside—what ripe tomatoes are like, or that the sky is blue. But what will happen when Mary is released from her black-and-white room or is given a color television monitor? Will she learn anything new or not? Philosophers used the analogy to debate whether experience is about more than simply having the knowledge of what creates that experience. When Jackson described the Mary's Room analogy in 1982, it was thoughtful speculation. Now Kim and I were living it and discovering the answers firsthand.

I learned that Kim too had grown up feeling like a social failure. But in spite of that shared background, the way we'd each internalized the TMS experience was strikingly different. Where I saw hope for the future, she saw an explanation for her former failure, and it was devastating. "Suddenly I understood why I have trouble with my friends, and why I don't get along with my co-workers. TMS showed me everything I'd done wrong in my life, and it overwhelmed me."

All of her observations made perfect sense, yet my interpretation of their meaning was different, and I tried to offer that as some comfort. I'm not sure if it helped, but it was the best I could do. Where I saw promise in my new understanding, she saw sadness in lost opportunities. Those are both real perspectives. Neither of us could ever go back and correct our failures. She was crushed, and for a moment, absorbing her feelings with my newfound sensitivity, I felt exactly the same. *Maybe she's right,* I thought. *Maybe this TMS has just shown us why we fail; maybe it's nothing more than a cruel joke with patients and researchers as unwitting participants.*

"What am I going to do now?" she asked. "It's like I'm haunted. I got a

---

* Mary's Room is described in Frank Jackson's 1982 article "Epiphenomenal Qualia." He expanded upon that in "What Mary Didn't Know" (1986). Other philosophers chimed in with their replies in *There's Something About Mary* (2004).

glimpse of those emotions but now it's gone. So now I know what life is like for other people but it's not that way for me." Her words made me question my own upbeat attitude. Had I only gotten a glimpse too? Maybe she was the one with a realistic viewpoint. I'd felt I'd acquired a superpower, but I had to wonder—was I just a smiling fool?

"I think I am still different," I said, but I had to admit that I wasn't sure. What I was sure of was that knowledge is power. Just knowing about the emotional brilliance that existed would forever shape how I dealt with people, even if I became oblivious to it again.

We formed a unique bond that night, talking about our different interpretations of a shared experience. Here's what she wrote about her experience, in a guest post on my blog, a few weeks later:

> I had an opportunity to participate in a study at Beth Israel Deaconess with Drs. Lindsay Oberman and Shirley Fecteau. They were using transcranial magnetic stimulation of certain brain areas that affect communication.
>
> I had a remarkable revelation during the study. I was able to see and hear what it is that ordinary people see and hear during a social interaction. I actually understood what emotions the facial expressions were conveying. I understood the tone of voice that the person was using, and what it was conveying. I also understood sarcasm, which I normally do not pick up.
>
> After seeing how differently my brain was working after the stimulation, it was clear that my brain usually focuses about 90% of my attention on the literal meaning of the words that are being spoken. Before this stimulation, I thought that I read people's facial expressions and their tone of voice fairly well. However, after seeing the difference following the stimulation, I would say that I miss 50% or more of a social conversation.
>
> If somebody says something sarcastically, I may completely miss his or her intention. My mind tends to focus on literal meaning. If they are being blatantly sarcastic, I can understand

that they don't intend the literal meaning of their words. However, my mind still tends to focus on the words that were said, even though I understand that's not their intention. In most cases, this feels quite uncomfortable.

If they are not as blatantly sarcastic, I think I sometimes pick up something in the way they are talking which doesn't sound quite right. The tone of voice doesn't quite match the literal meaning that my brain is interpreting. This also causes confusion and unease.

If someone is being sarcastic, but is speaking with a very straight face, I simply hear the literal meaning of the words and don't understand their intention at all. It seems to me that this is one of the main problems with a social interaction where someone is kidding me. I hear the literal meaning of the words, which sound as though the person does not like me, or does not wish me well. I don't hear the intention behind it, which is being conveyed by their face or their voice.

After seeing so clearly all of the aspects of communication that are conveyed by facial expression and tone of voice, I certainly understand why many people with Asperger's have social anxiety. In some ways, it is as if you are in a foreign country, and aren't completely fluent in the language. The people speaking to you don't know that you don't speak the language, and they expect you to understand what is being said. Your experience is one of trying to understand what is being said, and having to translate what is being said so that you can understand it. It can be anxiety provoking when people expect you to be able to react in a way that you are not able to do. There's certainly a lot that gets missed or misunderstood. It's also very tiring to have to constantly try to figure out what is being said, and to be wrong too much of the time.

I have great hope now that the researchers are finding some specific brain areas that are affected in autism spectrum disorders. As they are able to refine the stimulation techniques, they may be able to influence the brain in ways that can permanently improve communication. As I have just seen, this could be a life-changing experience for millions of people.

I like to think that our conversation that night helped Kim to see her experience in a more positive light. After getting over the initial shock of understanding her previous social failures, she too began to want that insight back. She also came to see knowledge as power, and both of us realized that seeing into people could strengthen us and change our lives.

The only question was, how to make it last.

Awakening

NEUROLOGISTS KNOW the brain's frontal lobe as the seat of higher consciousness, but exactly how that consciousness takes flight from a tangled mass of neurons remains a mystery. The frontal lobe is where we form ideas and make decisions, gather our speech, and consider what's right and wrong. It's also the center of what psychologists call executive function—in other words, organizing ourselves for daily life. Executive functions are things like getting dressed, planning out your day, predicting the consequences of something you might say or do, or deciding whether an action is appealing or not. The frontal lobe is also the region of the brain that interprets others' behaviors and shapes our responses to them. One way to characterize all that would be to say that it's the brain's center of abstract thought.

The frontal lobe includes a mass of connections that tie other brain regions together and a calculating engine that makes sense of it all. If the brain has a computational center, this is it. The frontal lobe also contains Broca's area, our speech center. Stimulating or suppressing parts of Broca's area with TMS can alter or take away the ability to speak. The target areas for my TMS study were associated with Broca's area too, but those stimulations had implications far beyond affecting speech.

Most early TMS work targeted brain areas with strong direct connections to the outside world. When those areas are stimulated, the results are perceptible right away. Stimulation of the visual cortex, for example, can make you see waves or sparkles of imaginary light. Stimulation of the motor cortex will make your fingers or toes move on their own. Those things happen the moment the TMS energy hits your brain. One early use of TMS was to help surgeons identify brain functions in those regions before they operated, and they still use it for that today. If you're going to remove a brain tumor, it's important to know what the brain matter around it does. TMS is still used on stroke patients, stimulating the motor cortex and other regions to help patients regain fuller use of their bodies. The immediate feedback from TMS stimulation makes the therapy possible.

Stimulation of the frontal lobe, in comparison, often has no visible effect because you're not stimulating brain cells with simple direct connections to the outside world. It had been easy for the scientists to figure out where to fire the TMS to make my index finger twitch. But knowing how to fire TMS into my frontal lobe to make me *want* to twitch my finger . . . that is infinitely more complex. There's just one set of neurons that actually moves my fingers, and their general location is well known to neurologists. There are a million sequences of thought that could set those cells into action, some of which are under my conscious control and others not.

Yet that's in essence what the scientists hoped to accomplish by beaming electromagnetic energy into the heart of my consciousness. The frontal lobe connections they targeted are thought to be unique to humans, and if that's true, these higher functions are part of what sets us apart from other animals. But the complexity of the frontal lobe meant that isolating what particular connections do is difficult. Any area they might stimulate would be connected to thousands of other brain areas, with most of those connections unknown except in the most general sense. So they could hope to guess correctly about the function of an area, but what else stimulating that area might do would have to be discovered by observation. That was my job, and I took it seriously.

After the study concluded I learned that the researchers had hoped that one of their stimulations would affect my ability to recognize words. They had theorized that autistic people might be slower to recognize certain words and that comprehension disability might contribute to emotional blindness. Even today I have no idea if that's true. I'm not even sure if they changed my word comprehension. All I know is that the unintended side effects overshadowed by a thousandfold anything they'd set out to test and measure.

The TMS surely had an immediate effect on the area where it was aimed. The laws of physics tell us energy was delivered there, even if we don't know exactly what it did. But that was only part of the TMS effect. The stimulation energy might well have stunned a group of brain cells underneath the coil and knocked them out of action for a few hours. That might be the suppression Alvaro talked about. Suppressing an area would produce effects the scientists could measure right after our sessions.

But suppressing one area might give other areas a new chance to communicate, and they might start using new pathways in my brain as a result. That process is what neurologists call brain plasticity, and it might well be encouraged by TMS. New paths, and old paths that are rendered accessible after the suppression of the target area, might do any number of things. And it might take a while for those new connections to build up enough traffic for me to notice. There would be no sign of impending mental upheaval as I sat in the lab doing the poststimulation testing. When I did notice, the effect was totally unexpected but still generally related to the original goals of the scientists. Interestingly, Lindsay later told me that I was the only participant who reported "delayed action effects," like my night of seeing music, or the hallucinations. Yet several participants in the study described changes in themselves very much like the heightened perceptions that came to me a day or so after stimulation. Did our timeline for experiencing results differ, or did I just describe the sequence of events more carefully? That remains a question to be answered.

Alvaro was one of the first researchers to stimulate the deeper cognitive brain regions. He'd been doing it for ten years before I came along, though most of his earlier work was with depression and other neurological problems, not autism. When we started the study, no one had really tried stimulating a small area in the "thoughtful" frontal lobe, or if they had, the results had not been published in a medical or scientific journal. I'd started looking at some of the journal articles and quickly learned that the number of TMS researchers in the world was quite small.

On one hand, it was neat to realize that Alvaro was one of the leading TMS experts and that he was breaking new ground with his exceptional team. On the other hand, the absence of complementary research meant we were on our own with no precedents or comparisons to guide us. I was patient number one for these autism-emotion experiments. At times it was thrilling to be on the razor's edge of research science, but at other times I felt like a lab mouse lost in a maze.

In my reading up on TMS, and contemplating its potential effects, I'd explored the history of brain stimulation. I had first thought "history" referred to work done thirty years ago, and I was shocked to discover that doctors were using electricity to stimulate human brains as early as 1860. That was when Eduard Hitzig, a Prussian army doctor, began zapping soldiers whose brains were exposed by gunshot wounds. Dr. Hitzig found that an electric current applied to the brains of these unfortunate victims—still alive but gravely wounded—made their muscles move involuntarily. The English scientist Michael Faraday had discovered something similar, a few years earlier, when he connected a battery to a frog's brain and saw it jump clean off his table. I suspect it was stories like these that inspired the fictional Frankenstein's monster, animated by the power of lightning.

Hitzig's ethical justification for electrifying the brains of wounded soldiers rather than treating their injuries has been lost to time, if it ever existed. Perhaps someone thought the electrification *was* a treatment. No one knows if the doctor made his patients better or accelerated their demise. All that remain are the papers describing his discoveries, and they're pretty disturbing to read today (especially while waiting for your own stimulation to begin).

After leaving the battlefield, Hitzig returned to Berlin and turned his experimentation to living dogs, applying electricity to different areas of their brains and observing the results. He used their responses to create a rough map of a dog's brain. His 1870 paper that describes the canine cortical map stands as the first real description of what we now call the motor cortex. The insights in the paper would guide a generation of scientists, but the method by which he gained this knowledge would trouble ethicists. Did the end really justify the means? Some people voiced their outrage at the notion of dissecting and experimenting on living animals. These protesters, who were known as antivivisectionists, banded together to challenge the medical ethics of the day. Thanks to them, the University of Berlin banned Hitzig's dog experiments from campus labs, but he continued them in a colleague's basement. Guys like Hitzig may have made great contributions to medicine, but they also inspired a thousand horror movies.

In London, neurologist John Hughlings Jackson was developing his own theories about the functions of the nervous system, including the then-novel idea that epilepsy originated in the cortex. He'd formed that opinion working with patients at the National Hospital for Paralysis and Epilepsy, and he passed his thoughts on to David Ferrier, a younger colleague. Ferrier was already fascinated by Hitzig's work, and he decided to test Jackson's theories. By 1873 he was inducing epileptic seizures in dogs by using a battery to stimulate the exposed cerebral cortex. Just as Hitzig had mapped out the motor areas, Ferrier mapped out locations on both sides of the brain that induced seizures.

Ferrier and, presumably, the others were well aware that cutting open the skull of a living creature was painful, and he suspected that electrical stimulation might hurt too. The problem, as Ferrier described it in an article published the following year, was that anesthesia made his subjects less responsive to stimulation. Without anesthesia, however, responses would be more difficult to interpret because they might be the result of stimulation, reflexive pain responses, or attempts to escape. Ferrier, Fritsch (another nineteenth-century brain researcher), and Hertzig had all described tranquilized animals crying out in pain after some stimulations. For those reasons, and what he called "the sake of humanity," Ferrier used anesthesia "before and throughout" his experiments.

However, that wasn't enough to diffuse the public's horror and outrage when word leaked out about his research. There was a tremendous outcry by antivivisectionists, but the importance of his work ultimately proved to be his defense. His 1874 paper and the work that followed established Ferrier as one of the great experimental neurologists, and he was elected a Fellow of the Royal Society three years later.

Once that paper came out, it took less than two years for someone to try systematic electrical stimulation on a living human brain. Roberts Bartholow* was a well-respected physician at Cincinnati's Medical College of Ohio. He believed first-person experimentation was the best way to advance knowledge of the brain. When he found himself presented with a rare opportunity—a thirty-year-old cancer patient whose skull had been perforated by a terminal cancerous ulcer—he was quick to act. There's no record of how he went about getting the patient's permission, or what he may have promised her, but shortly after making her acquaintance, he was prodding the hole in Mary Rafferty's head with several needles connected to a primitive but powerful battery.

According to Dr. Bartholow, Rafferty was friendly and "eager to please." Bartholow assures his readers that he had her "full informed consent," but just a few pages later, he describes her as "feeble-minded." His descriptions of the experiments were even more disturbing. After inserting a wire into her brain, "she complained of acute pain," and his response was to increase the current! At that, "she showed great distress and began to cry." Soon after, she frothed at the mouth and passed out.

Dr. Bartholow's report remains in the archives of the National Institutes of Health, where you can read it today. In it, he describes what came

* Harris, L. J., and Almerigi, J. B. "Probing the Human Brain with Stimulating Electrodes: The Story of Roberts Bartholow's (1874) Experiment on Mary Rafferty." *Brain and Cognition* 70, no. 1 (2009): 92–115.

next. "After spending a day in bed, [she was] decidedly worse . . . stupid and incoherent." Things didn't get better with time. Rafferty had a seizure and slipped into a coma following a second round of electrification. She died a few days later, and Dr. Bartholow removed her brain and dissected it during the autopsy. He published an account of her case a few months later.

There was some outcry in the press, and one doctor in town sent a complaint to the judicial council of the American Medical Association, but little came of it. A few years later, in response to continued professional criticism, Bartholow expressed regret that he had caused Ms. Rafferty some injury and pain, though he also said she was "hopelessly diseased" and cancer already "threatened an early extinction of life."

"She would have died anyway" is no longer a justification for this kind of mistreatment, and the standards of medical ethics and informed consent in research are very different today. Still, it's a wonder that after reading all that grisly history, I managed to stay so upbeat. It's a testament to how strong the promise of life-changing discovery can be.

When I looked back at Bartholow's work from the perspective of today, I felt shock and horror. Yet he was a renowned and widely admired brain expert by the standards of the late nineteenth century. Who's to say that in a hundred years a scientist might not look at the experiments in Alvaro's lab much the same way? I always felt absolutely confident that Alvaro did his best to look out for me (and everyone else in the study), but his ability to do so would always be constrained by the knowledge of the day.

And with that in mind, I hadn't forgotten Alvaro's gentle warning: "The effects you feel might not always be so positive." We talked about this at some length the next chance I got to stop by his office. Busy as he was, he always had time to consider my questions, often with no notice. We'd sit at his conference table, with tall bookshelves full of books about neuroscience surrounding us and light streaming through the window in the corner above his desk.

Our talks were always enjoyable. I could never predict what he would say. Sometimes he spoke of contemporary research, but he was just as likely to bring up the ideas of some philosopher or scientist from one hundred or two hundred years ago and ask what that might tell us about our current question. That day, he seemed confident that there might be a single interesting explanation to account for why my changes were mostly good, and why they seemed so durable. "Remember that sled-run analogy that we talked about, and how a path becomes more natural and easier to use the more times you run it? Maybe that explains what happens here. When we do a stimulation and it produces an effect you don't like, one that makes you uncomfortable, you don't choose to run that path. But when we get a result you *do* like, you ask your friends if they see any changes, you look for them yourself, and you run the path as hard as you can. So your own mind is selecting and reinforcing for a good result. Maybe nothing protects you from having a bad effect in the initial stimulation, but your mind essentially throws the bad away and focuses on what it thinks is good."

I still wasn't sure if his theory explained the way changes were settling in for me. "I had this really incredible sense of seeing into people and being part of the music. That insight went away quickly, but a new sense of emotion seems to be growing in its place."

"Awakening." Alvaro smiled a bit at that thought. "That seems to be what's happening. This sense of emotion is awakening in you. You've always had the ability to feel deeply—that's clear from your writing. But what's awakening in you is the ability to feel deeply in response to what you see in others, or what you hear about them." The implication of his words was that the emotion I was feeling now in response to things I heard or read was an ordinary human ability or trait, but it had been suppressed or asleep in me all these years.

I'd started to realize that there was a continuum, with highly logical thinking on one side and emotional response on the other, with most of humanity somewhere in the middle. Alvaro agreed with that, but he added that our position on that continuum is a dynamic thing, as the brain regulates our responses to suit the environment. "It was biased toward the rational end for you," he added, "and maybe now it's not so much.

"Yes," he continued, "that sounds right. I think the wiring to do these things was in your head all along. That's the only way they could have come to life immediately after TMS. Perhaps you never knew how to access them, or perhaps the circuits were suppressed and the TMS was able to bring them to life. As surprised as you were that the effects didn't come to you till the next day, that's nowhere near enough time for your brain to build new pathways. So they must have been there."

He sat there, thinking for a moment. "When we do the TMS for depression, that has a delayed response too. Some patients say they walk out feeling better, but for many of them, it builds over time, somewhat like you reported."

His words made me wonder what else might be prewired in my brain. Did we have multiple personalities all wired up and ready to use, just as "emotional vision" was always latent in my mind, waiting for activation? Thirty-five years earlier, when I was fifteen, my mother's terrifying hallucinations had been totally real, just as real as my sense of seeing into other people. "The monsters are right up there, John Elder. Two of them," she would say during her psychotic episodes. I'd look where she pointed but all I could see was the ceiling. Still, she was so sure there were demons that she became angry when I could not see them.

The doctors at the hospital tried to reassure me by saying there were no demons and it was "just my mother's psychosis talking." They would medicate her, and she would emerge in a subdued but normal state a few days later, but I never forgot the speed with which she threw a switch in her mind and began seeing a totally different reality than any I had known. Ever since then I had wondered if demons are inside all of us, waiting for the right trigger to set them free.

"TMS may be particularly effective for you because you have an exceptional power for self-reflection. Maybe that's why you wrote your book, or maybe you developed that ability by writing. Either way, it's uncommon and it may help you as you identify a change and focus on it, whereas someone who is less internally aware might overlook it."

That raised an interesting possibility. Alvaro had said that some of the younger participants in the study were not noticing any effects at all, even though they might show differences on the tests. I didn't meet those other volunteers, and they remained anonymous to me. Still, we speculated about why I described experiences that they didn't report. Was it necessary to notice the effects of TMS and "work the changes" to get an effect? Might my own efforts to do that have accounted for the bigger changes in me?

"It's possible," Alvaro said. "Perhaps the TMS would be most successful if it was done in conjunction with therapy. But there's a problem with doing therapy in a study like this. Critics would ask if it was the TMS or the talk therapy that made the difference, and it would be hard to separate them."

I realized that was true. From his perspective as a researcher he wanted to keep the experiments simple. But from where I stood—as an individual who wanted to change—I could definitely see how counseling or discussion might make all the difference.

That first TMS experience, the awakening of my insight into music, had taken me by surprise. No one had predicted such a result, which told me I should be open to just about anything that might come next. So as the sessions continued, I think I paid closer-than-average attention to my reactions. I knew that greater emotional intelligence was the general target, but I had no idea exactly what to expect. The fact that the researchers didn't know either told me I should take special note of anything out of the ordinary, because I had no idea what would be important and what would be irrelevant.

I contrasted my experience with the little I knew of some of the others in our study. Some of them lived alone and didn't interact with as many people as I did. Some of them may well have been temporarily changed by TMS, but then maybe they went home and were alone all night till the effects wore off. In that case, they may have experienced some of the same changes as I did only to have them fade away, unnoticed.

My son had felt a change in his perceptions, but he didn't ruminate on what happened as intently as I did. He seemed to alternate between uninterest and excitement when it came to TMS. Still, he never forgot the sense of sharpening the TMS gave him, and both of us wonder how much permanent effect there was on him. It's hard to say because he wasn't thinking much about people and emotional insight when we did the study. He'd wanted a distraction from the upcoming trial, and his interests remained technical—chemistry and computers. TMS hadn't changed those things in his mind. It's also possible that teenage boys simply aren't emotionally mature enough to be so self-reflective. I could see—with considerable unease—that the foundations of my own marriage were shaken by TMS, and I had no idea if anything similar was happening with my son. I wondered if TMS had affected his relationship with his girlfriend, but it was his first romantic relationship and I reasoned that he had nothing to compare it to. Seeing the world differently is a really big deal—all the balances in your life change. Martha's fears had felt like depression talk in the beginning, but they were proving prescient as the study unfolded. My friend Michael Wilcox didn't think he'd been hit with a new sensitivity to emotion when we discussed it right after the study, but now he feels differently.

Michael recently told me that he's become unable to read fiction, and that came to pass after TMS. He had always eagerly awaited the weekly arrival of *The New Yorker* magazine and the terrific new short story in each issue. He had once dreamed of writing himself and had taken creative writing classes back in his college days at Harvard in the early 1990s. But after TMS, his love of fiction was replaced by an absolute dread of the emotions that would well up in him when he read of others' misfortunes. And that remains true to this day. It's funny, but neither of us shared our newfound emotional fragility until we spoke this past fall, and each of us was surprised to hear that it had occurred in the other. I think we were both a little self-conscious and embarrassed at the notion of grown men reduced to tears by reading magazines!

What about the hallucinations I experienced? None of the researchers had a ready explanation for them. Dr. Timothy Leary—another Harvard faculty member—had described his drug-induced hallucinations as visions that facilitated life changes. My own life sure felt different since they had appeared. Dr. Leary had studied hallucinations that came from psilocybin mushrooms, like those I'd recalled that night in the pizza parlor.

I thought about friends who'd done mushrooms or LSD back in those days, and how five of them could eat the same mushrooms while only one or two described having hallucinations. Some of them even complained about the lack of effect. "Nothing's happening," one would say, and he'd eat another handful. At that point things might explode, but one fellow's "handful dosage" might be ten times what it took for the most sensitive fellow in the group to be sent into orbit.

Perhaps something similar happened with TMS. Earlier life experience had shown that I might be particularly sensitive, and perhaps everyone else would have shared my experience had their dose been doubled or tripled. I also wasn't sure how many participants would speak up, even if they had the hallucinations I'd experienced. Many people would be self-conscious or embarrassed, given the status of hallucinations in today's world. The sixties and seventies were a very different time.

"What makes you think you are unusually sensitive?" That seemed a reasonable question for Alvaro to ask me when I raised that possibility. Luckily I had a better answer than a long-ago psychedelic drug experience. I told him about three recent occasions on which I'd been given prescriptions for psychiatric medication (for depression and anxiety—both of which are common in autistic people), and how each time I was overwhelmed by what the doctors described as "minimal starting doses." With each medication, I ended up taking a fifth to a half the standard dose, and I discontinued those quickly because I felt they still affected me too much.

Lindsay wasn't surprised at all. Several times she had remarked on how I sensed things that no one else commented on—like counting the multiple taps of a TMS burst, or describing the stirrings inside my head.

In recent years there's been some research that validates my experience—we now know autistic people can have exceptional sensitivities, such as mine for sound. And other autistic people do seem extraordinarily responsive to psychiatric medications. But all I knew for sure then was that some guys could drink a twelve-pack of beer and not show any effect, whereas if I drank one can, I felt a buzz. That was one reason I stuck to iced tea for refreshment and stayed totally clear of illegal drugs—I'm too old to be getting high and flying my motorcycle to Mexico.

Or maybe I'm a little bit scared to find out what would happen if I actually went there. The more I reflected, the more I understood what powerful and prophetic things hallucinations could be. Back in the day, most people dismissed them as just being high. But I knew others who felt themselves powerfully changed by them, even years later. As an interesting side note, other researchers are now exploring the use of hallucinogens to help terminally ill patients find peace and comfort through expanded consciousness. Perhaps the TMS experiences gave a little of that to some of us.

Science Fiction
Becomes Real

WAY BACK IN EIGHTH GRADE I read the science fiction novel *Flowers for Algernon*.* If you've read that book or seen the movie, you probably remember how things begin so hopeful and end up so tragically for Charlie, the main character. He starts out as a cognitively crippled janitor in a bakery, and then some scientists change his brain and turn him into a supergenius. In a matter of months he goes from cleaning toilets to writing cutting-edge research papers. But then Charlie discovers a flaw in the science that made him smart, and he watches his intellect evaporate as fast as it arrived. I suddenly found myself haunted by the story, and wondering if I wasn't a bit like Charlie. My own transition was nowhere near as dramatic as his, but the TMS study still marked a tipping point in my life.

I'd read a lot of science fiction, and I'd always been amazed by how prescient some of my favorite writers could be. Now I found myself playing a role in one of the stories, and I was chilled by the notion that it

* *Flowers for Algernon* was the basis for the Academy Award–winning 1968 movie *Charly,* starring Cliff Robertson.

might predict my future. The parallels became more and more apparent as the study and its aftermath unfolded.

Before TMS I was a car mechanic in a small New England town. After TMS I emerged on an international stage, sought out for my thoughts on autism and neurodiversity. It felt like I'd learned more about autism and brain science in the last few months than I had learned about other things over the past fifty years. And I was soaking up new knowledge as fast as I could gather it, by reading books and articles and absorbing every bit of wisdom I could from the doctors and scientists around me. Some would say my first book had precipitated these changes, and indeed its publication opened many doors. But I believe the insights I gained from TMS—and some new abilities—were what allowed me to walk through the doors and seize the opportunities.

Friends and even acquaintances had started to say I seemed different, though they weren't sure exactly how. They'd said that to Charlie too, when he'd started to change. But how long would it last? If *Flowers for Algernon* was any guide, it was time to start worrying. I'd seen TMS turn on latent abilities in me, only to watch them fade away hours or days later. I was never sure what essence remained, or what had been lost, and I wondered if I would end up like Charlie.

There were moments when I felt I'd gained so much . . . and that convinced me there had to be a price to pay. There were times I'd wish that I'd never started this TMS, because I'd been functional the way I was, even if my sense of other people was weak. Before I began, it had never occurred to me that I could lose what sensitivity I'd started with. Thinking of *Flowers for Algernon* had me worried that TMS was like a poker game, and that these brain experiments could end up cleaning me out.

That thought became harder to ignore as some of my newfound powers slipped away just as I was getting to know them. When I learned I was autistic, in the late 1990s, the therapist who'd first told me said, "You were born this way," and indeed I'd never known anything else. Nor had I ever questioned it. I'd always made the best of things, and after learning my diagnosis I taught myself how to act in the ways others expected, in order to fit into social settings. It didn't always go smoothly, but it helped me find some stability and some professional success. But inside, I was the same as I'd always been—or so I thought. It was Alvaro and the other scientists who showed me how limited that viewpoint was.

Psychiatrists, psychologists, and neurologists have evaluated me on several occasions since that first diagnosis. Each time, before they began, I'd feel that old fear of being exposed as a fraud. You might be surprised I'd feel that way about something that was called a disability diagnosis. But it was true. After a lifetime of feeling defective, it was far better to learn that I was a typical autistic person and not a defective human. The prospect that the diagnosis was wrong always scared me, because that would mean I was just defective. It didn't matter how much money or success I attained—that worry never left me.

It was the same feeling I'd get in grade school, when the answer to a math problem popped into my head and my teacher said it didn't count because I couldn't show my work. Or much later, when I lied about my credentials to gain employment as an engineer. But every time, the diagnosis remained the same. I start every evaluation fearful that I won't measure up to some imaginary standard, and then I'm disappointed that I'm not instantly better, even though I've always been told that autism will never leave me. Yet Alvaro, Lindsay, and their team gave me hope. "Our brains are changing all the time, rewiring themselves through the process we call plasticity," Alvaro said. "You practice math and get better at it, right? So it should make sense that you'd practice social interactions and get better at them too. Both involve fine-tuning the brain to do things more efficiently."

"We think TMS may help and accelerate this fine-tuning process," Lindsay said.

There was a time when I would have preferred to do my fine-tuning in private, but that moment was long past. Sharing my stories with the world in *Look Me in the Eye* showed me how much interest there was in autism and in emerging from disability. Consequently, I felt a sense of duty to my readers to keep writing about my journey. That first night at dinner with Alvaro and Lindsay, I'd become so excited about the possibilities of TMS that I'd asked them whether it was okay to share the news on my blog.

"Sure," Lindsay told me. "Describe what we are doing and let's see if anyone volunteers for the study." Lindsay pointed me to the description she'd put on a flyer. "We have to be careful what we say," she explained, "because the hospital's review board keeps a close eye on what we tell potential research subjects. It's important not to mislead them or promise things we can't deliver."

When I originally posted my description I did my best to make the limited goals of the study clear, but a surprising number of readers still took my words to mean that TMS was a ready-to-go therapy for autism, even though I expressly said it wasn't. When I looked at the traffic statistics for my blog a few days after that first post, I was shocked to discover the story had been downloaded several thousand times. Beyond that, Lindsay was swamped with a hundred-plus phone calls and messages to answer. When the scientists had first reached out to me they were having a hard time finding adults with autism for their study. But after I'd written a few blog essays on TMS, volunteers were coming out of the woodwork.

"We've had four new people join the study this week," Lindsay told me excitedly, and the project had barely begun. The realization that I'd played a small part in that made me proud. As people continued to sign up I just hoped I'd made a good choice by talking publicly about the research and that the people who followed me into the depths of TMS would be helped and not harmed by the experience.

That had sounded great when I'd acquired new insights at every visit to the lab, but what was I supposed to say when the gifts I celebrated slipped quietly away? Alvaro and Lindsay both reassured me on that score. "First of all," Lindsay told me, "there is no reason to think you will lose the intellect you had before TMS just because you can't look deeply into someone's eyes anymore. We told you at the outset that any gains from the study would be temporary. What you have described is still remarkable, wherever it goes from here."

Perhaps my own sense of inferiority—which I attributed to my autism—had made me overexuberant when I initially described its potential. After all, I wasn't a neurologist. "It's great to get volunteers," Lindsay had said, "but I'm worried that callers are expecting too much. They are looking for treatment, and this isn't a treatment study. This is just basic research. Quite a few of the people who call me don't understand that."

Her words told me that there were a lot of people out there like me who have muddled along with some level of disability but were ready to jump at a fix the moment it was available. Then there were the mothers who called Lindsay looking for help for their children. Hearing about them made me sad, because I knew how lonely I had been as a kid, but I also realized that autism had contributed to my success in other areas as an adult. Any mother looking at me as a boy would have been in a desperate panic for sure.

"Being lonely as a kid might well have been necessary for me," I told audiences in my talks. "If I'd had the friends I dreamt of, I'd never have spent the time to become the machine aficionado I am today. Now that I'm grown I can put that in perspective. The world is full of friendly people with no technical skills. The few of us who see into machines like others see into humans are singularly uncommon, and we're valued for that. If we use a technology like TMS to help a lonely teen today, will we be taking that exceptional ability away from him tomorrow? Should we trade friends in seventh grade for designing a working spaceship at age twenty-five?"

Those were hypothetical questions in 2008, but I believed they would be real issues soon, and I wanted to start a dialogue. Unfortunately, many of the parents who called Lindsay weren't interested in talk. They wanted a fix, and they wanted it now. Neither of us knew how to respond, except to tell them this was basic science and not the miracle cure they were hoping for.

The thing I never revealed was that I had as much hope as the most enthusiastic of her callers. Attitude means a great deal when you are sick; studies have shown that patients who believe they are getting better recover at higher rates than those who give in to their illnesses. Autism isn't a sickness, but I hoped the analogy would hold and my positive outlook would translate into real success. In those moments when I questioned myself, it was encouraging to recall that I'd been in a similar position thirty years earlier when I told the guys in KISS that I could definitely make a guitar blow fire. "No problem," I assured them, even though I'd never done anything remotely like that before. With my foolish promise and their blind faith and hope, we made it work, and every guitar I did for them was a success.

Yet my online postings about the program also drew lots of questions and a wide range of commentary. Many readers were dubious, skeptical, or both.

One of the hardest questions to answer was "Why are you doing this?" After writing about all that I'd done to make a good life, people found it hard to understand why I would be so quick to embrace change. That was tough, because I didn't really understand it myself.

Then there were the technical questions: What area are they stimulating? What protocols are they using? It was intimidating, and I resolved to acquire the knowledge I needed to get it right. I had succeeded till now with this method, often starting with nothing. After all, by the time that series of smoking, burning, and illuminated musical instruments for KISS was done, I was surely the top expert anywhere on the construction of fire-breathing guitars, even though I had started from nothing. Who else was even dreaming about such a thing in the 1970s?

This situation seemed simpler, at first. All I had to do was learn how to explain the TMS process. Unfortunately, explaining TMS turned out to be like peeling an onion. Each layer of explanation revealed another question, and I soon found myself at the limits of medical knowledge.

Whenever I thought I was ready, I wasn't. The questions came pouring in, and though I've sometimes said that I learn by explaining, when I started explaining TMS I realized that my knowledge of electricity, engineering, and physics was not enough. The simplest of questions proved devilishly complex.

What does it do?

How do you know it's safe?

Or take the most basic question: *How does it work?*

The answer begins with the transference of electromagnetic energy from the TMS coil into the microscopic threads inside the brain. But as I write that sentence I can already see the questions forming. Why? How does it do that? What are threads of the brain? My first answer was enough for some people, but others wanted to know more. Including me. The only solution was to follow the energy down the TMS rabbit hole in the hope of emerging with more solid knowledge. As I kept learning, I kept writing, and the questions kept coming.

What does the energy do when it gets into the neurons?

To answer that I turned first to the neuroscience texts Alvaro had recommended, and when I wasn't fully satisfied, I turned to the maestro himself. I'd come to believe he could answer anything, and I was somewhat startled to find that his knowledge had limits too. Still, between his insight, the discussions with his postdocs, and a lot of reading I began piecing things together. The further I ventured into the brain cells, the more complex things got.

The brain makes its electricity by breaking down enzymes that are delivered by the bloodstream along with oxygen and other essential elements. That electricity forms signals that get passed from neuron to neuron, just as electricity in a computer passes from one component to another. With a computer, you press a key and a signal is sent to the processor, which sends signals to a printer, where a document appears in the print tray. In a brain the eye sees some bacon, and the nose confirms the odor. The brain sends a signal to the arm, which grabs the morsel and puts it in the mouth. What do those things have in common? Both are sequences controlled by tiny electrical signals. How might additional energy affect those processes?

At first glance, it seemed like TMS would scramble the circuits. "That may explain what's happening," Alvaro confirmed when I asked him. "TMS might be like a reset button for the part of the mind where it's aimed." But then he continued to explain that TMS could be "turning things up," just as jacking up the power in a stereo makes the whole thing get louder. Another possibility is that TMS is changing the network in our heads by temporarily rearranging the wiring. The plan being followed in Alvaro's lab was essentially: pick some possible targets, stimulate them, and see what happens.

As I learned more, I began to see that much of research medicine operates this way. Until quite recently, we had no idea how drugs worked. We discovered them by chance and experiment and used what worked. Penicillin is a good example—it was discovered in a lab culture in 1928 and immediately put into use. Its chemical structure wasn't unraveled until sixteen years later, and it would be decades before we'd truly understand what made it work.

We appear to be in a similar place with TMS—we know it's got power, and it has certain effects, but we're far from fully grasping the fine subtleties of its operation. But that doesn't mean we can't use it to great benefit as we unravel the science behind it. This learning stage was a humbling experience for me. To think I'd imagined that I might have an advantage over the doctors with my grasp of physics and engineering! Now I saw that the little I knew only took me deeper into a maze where there were no certain answers.

As I immersed myself in neuroscience tomes, more and more people were reading my book and blog, and I found myself answering fresh emails every morning. Some readers even called me on the phone. It always surprised me when the phone rang and it was an autism caller, as opposed to someone with a broken Mercedes or Land Rover.

I had intended to write about the promise of science and how an exciting new tool could potentially help autistic people. When readers called me, though, they were not thinking research. They were thinking treatment, and where they could get it. I found myself in a strange situation—I was a car mechanic explaining cutting-edge neuroscience to strangers who listened to my words as if I were some kind of expert. The best I could do was to describe the engineering principles and leave the brain part to the doctors. I referred a lot of callers to Lindsay at the lab and studied as hard as I could.

There were also a few calls from parents looking to apprentice their kids at my repair shop, Robison Service, in the hope that they'd learn the automobile trade. While that sounded great in theory, there was really no practical way to do it, though I wished we could. The idea of a roomful of autistic apprentices slaving away appealed to the capitalist in me, but it would have been an overwhelming task to train and supervise them all. Yet four years later, we finally did make use of those willing apprentices. A psychologist with a broken Jaguar introduced me to the folks at Tri-County Schools, a local program that serves public school students with developmental challenges. They were looking for ways to teach their students real-world skills so they could be self-supporting on graduation, and in the summer of 2013 we joined forces with Tri-County and opened a special-ed high school program in the Robison Service car complex. As I write this book we are in our second year of operation, doing exactly what those early callers hoped for. High school seniors learn real-life skills in our program, and they get to work as paid apprentices, overseen by the school's teaching staff. We're even using some of the neuroscience insight I gained these past few years to teach them. Our school psychologist uses brain-wave monitors and neurofeedback to help students learn to regulate their behaviors, and we are testing self-regulation techniques developed in Dr. Minshew's lab. So far, it's all working. The idea that I'd do something like that would have been inconceivable to me before TMS.

As I cast my autism writing out into the world, calls also came in from other scientists and researchers. In the summer of 2008 I began getting calls from nonprofits who wanted my input on autism research directions. They invited me to work with them, to select promising autism research for funding. Then I got a call from the Office of Autism Research Coordination at the National Institutes of Health. "We're conducting a review of proposed autism research," the woman on the line told me, "and we wonder if you'd be willing to help us out." My first thought was *Why me?* I'm not a doctor, and I have no experience evaluating research. Yet the caller ID said she was from the government, so I listened to her closely.

"We want to bring community members into the research selection process," she told me. "We've found parents to serve on the committee, but we've had a hard time finding adults on the autism spectrum who are willing and able to serve. We think your input would be valuable. Will you join us?"

I decided to give it a shot. I'd be asked to read research proposals and cast my vote for how the proposed science would affect the autistic community. I was supposed to use my judgment as an autistic adult to decide which pieces of research would be most relevant and beneficial. My opinions would be weighed along with the opinions of parents and autism scientists.

My friends were all supportive of the idea. Dave said, "That sounds really cool. I'd love the chance to do something like that." All that had me feeling pretty good, and I wrote on my blog about my decision to join a review board. The response from my readers was decidedly more mixed. "They just want to use you," skeptics warned. "You'll be their token autistic." I could not decide if I should feel offended or sad. Some comments were downright angry. "Why would they pick you?" one writer ranted. "You're a high school dropout with no qualifications to review autism science." "He's just jealous," my friend Bob reassured me, but those critical reactions troubled me a lot.

Martha wasn't so enthusiastic. She liked the idea of a trip to Washington, but she worried that my interest in autism science was taking me away from the livelihood that supported us. The science itself wasn't of any interest to her, perhaps because it scared her. Looking back, I realize her concerns were real and valid, but at the time it drove one more wedge between us. Perhaps she just didn't understand how important autism research had become to me, or how fast I was changing inside.

The review process itself was fascinating. Our group was to review about 155 proposals. I had been assigned 6 of them, but being inquisitive, I actually looked at all 155. We gathered together in a D.C. conference room, where we discussed and voted on the ideas.

Each of us had given scores to the proposals we were assigned, and those scores were tabulated so that the proposals under discussion were ranked from best to worst. We then discussed every proposal, and each of us had an opportunity to comment or question. I took full advantage of that, asking many questions and offering almost as many comments. By lunchtime I began feeling insecure, wondering if I was talking too much or in the wrong tone of voice.

TMS may well have sensitized me to that insecurity. In my earlier days of working in corporations, my commentary during meetings often got everyone upset, even when I was right. The way I saw it, I was just doing my job. Better to tell a roomful of engineers that their design was no good, and that they needed to redo it, than to have to explain it to the chairman of the board when a million flawed widgets were going up in smoke out in the field. My keen technical insight and direct communication style won me the opposite of fame and admiration in the corporate world, and most of the time I had no clue what went wrong. All I knew was that they didn't want me around anymore. These science reviews were shaping up differently. There were a few proposals I was opposed to, and a number I didn't support. But I didn't call any of them drivel, and I gave everyone a chance to speak without calling them fools.

Scientists approached me during our breaks to talk and exchange pleasantries. They asked me questions, offered comments, and showed genuine interest in what I had to say. I wouldn't credit TMS directly for making such dramatic changes in me, which in turn made those encounters successful. Rather, I believe it opened my eyes, and with my expanded vision and insight, my brain changed itself so that I fit better into that group.

The government moderator running the meeting affirmed that I was welcome there. "Those are good questions you're asking. Keep it up." Some of my words even sparked lengthy discussions.

After we were done the chairperson asked if I'd be willing to serve on other committees. "We could use more of your insights," she told me. I was proud to have done a good job and honored that she'd asked. As I pictured my staff putting water pumps on Land Rovers the next day, I thought of the phrase "cognitive dissonance," and I wondered if I was facing a similar mental situation now, having to function effectively in two different worlds.

The early committee service led to other appointments and my increased involvement in autism science. My world was becoming so different, so fast. I was proud of my contribution, and of acceptance in a new world, but I was often terrified and also lonely. For the first time in quite a few years, Martha was not beside me in a new field of endeavor. Some people seemed to go it alone as a way of life, but I wasn't one of them. For me, the changing world and the absence of someone to turn to was very frightening.

This science was fun while I was immersed in it, but when I got home and lay in bed at night and my mind started to wander, it got really scary. My friends thought my new life was cool and exciting. No one considered that the sense of stability I'd built so carefully over the years was now totally gone. Before I wrote a book and got into this TMS study, I just woke up in the morning and went to work, like millions of other people, doing the same things I'd done for years. But now those days were gone, and there was no telling what tomorrow would bring.

The researchers loved the uncertainty of what was unfolding. Lindsay said, "That's the best part of my job! Every day is different, interesting, and exciting!" I smiled and agreed when I was in the lab with them, but when I returned home at night, I had to work hard to keep from feeling terrified.

My connection to Martha had changed in a fundamental way, as I found myself soaking up depressed feelings that had formerly rolled right off me. How was I to deal with that? She told me I had changed and things weren't the same between us. She was scared, and so was I, because I believed she was right. *I had changed.*

After many years of accepting how she was, I found myself thinking I couldn't handle her depression. She in turn interpreted my newfound vulnerability as a rejection of her, and that made me sad and confused.

For the first time, I was absorbing fear from people around me. The emotions swirling around my son and his upcoming trial were awful. That was a big and unwelcome change from my former Mr. Spock self. Looming over it all was the *Flowers for Algernon* fear—the possibility that I might experience other wonderful things, only to see them slip away to leave me more disabled than before, and now fully aware of what I lacked.

Every night, when the excitement of the day had faded, fear remained. All the bad emotions around me were scary. The possibility that my marriage was failing was horrible. The thought of my son in jail was terrifying. And the idea that I might have set all that in motion through a misguided sense of inferiority and a desire to make myself better was very unsettling. But the die had been cast the moment I'd walked into the TMS lab. As Dr. Hunter S. Thompson said, "Buy the ticket, take the ride."

But Thompson shot himself in 2005, and *Flowers for Algernon*'s Charlie died in an institution. The excitement I'd felt with new TMS experiences was great, but I was now experiencing deeper low periods than I'd previously known in my life. I sure didn't want to end up like those two, but I didn't know how much power I had over my fate. The Thompson quote felt disturbingly apt—my emotions were taking me for a ride, and all I could do was see where it led.

The Zero-Sum Game

WE'VE ALL HEARD this myth: humans only use 10 percent of our brainpower. Usually, when people say that, they are suggesting that if we could learn to use the idle 90 percent we would become intellectual giants. Various supplements and therapies have been hawked over the years in pursuit of this lofty goal, but none of them has turned out to enrich minds, though I'm sure a few enriched their marketers.

The notion of wasted brainpower has been with us a long time. It seems to have started with a medical man and not a snake oil salesman, though one hundred years ago those two were sometimes one and the same. In "The Energies of Men," published in 1907, psychologist William James wrote, "We are making use of only a small part of our possible mental and physical resources." Unfortunately, Dr. James was wrong. We have no way of knowing exactly what he was thinking when he wrote those words, but everything we've learned in the hundred-some years since tells us there are no extra, unused pieces in the brain. We may not use all the parts of our brains in the most effective ways, but they are far from being idle.

We've known for a long time that any brain damage will cause some kind of functional consequence. We may survive the injury, but we are changed. That alone should be enough to tell us that the brain doesn't have any wasted space. Yet it is true that we don't use the whole brain all the time. Still, we use more than many people imagine. You might think, for example, that we don't use the part that moves our muscles as we sit in a chair and solve math problems. But Lindsay pointed out that wasn't true. "You may think you're sitting still," she told me, "but your brain is sensing your position and sending thousands of signals to muscles in your body and your legs. They are constantly making tiny adjustments to keep you in balance. That's why you'll fall out of a chair if you pass out— because that stops." During the course of a day every part of your brain will be called upon to do one thing or another. The parts that do particular tasks repeatedly get optimized for the jobs they do, which is how we become proficient. Some brains are optimized for cerebral things, like the study of history. Other brains guide their bodies to athletic excellence. Most brains—like most people—have a more generalist orientation.

A century ago, we didn't understand what certain brain areas did—and we still don't in some cases—so we assumed they did nothing. If a brain region didn't twitch a limb or cause a yelp when electrified, early neurologists wrote the area off as unknown territory in the best of cases and empty space the rest of the time. It's easier to say "It doesn't do anything" than to admit the truth, which was that they had no idea how to figure out the inner complexities of the mind.

New brain imaging tools like functional MRI allow neurologists to see activity across the brain as we do tasks, and we see how the whole brain "lights up" even as we do the activities associated with ordinary living. We still may not know what certain areas deep inside our heads are doing when we solve a puzzle, but we at least know they are somehow part of the solution. That's the first step to unraveling that mystery. We've also learned that the brain is constantly changing, refining its connections, and optimizing itself for the world around us. In fact, we recognize a "use it or lose it" mechanism at work during brain development. Our growing neurons form interconnections as we acquire skills, and those that don't make useful connections die off and are absorbed by the body.

This pruning process is now seen as an essential part of human development. I pondered what it might mean for someone like me, as I was influenced by TMS. In particular, I wondered what happened in my mind when I developed new abilities after several stimulations. Did I truly get smarter, only to revert to my old self when the effect wore off, or did I temporarily trade sharpness in one area for brainpower somewhere else, recognizing what I'd gained without seeing what I'd lost? It's sometimes hard to avoid a computer analogy when considering brain function, and I thought to myself, if my whole brain were already in use, it would be like a computer running at 100 percent load. You could add a new task, but the other programs would have to slow down to make room. If that were true, I must have lost something somewhere to experience greater insight into people.

Alvaro wasn't so sure he agreed with my line of reasoning. "Remember that the brain is constantly rewiring itself—something computers cannot yet do. The number of brain cells may stay the same, but the connections between them change. That process of optimization happens with practice, and it's how you get good at things. Learning to solve math problems faster doesn't seem to come at a cost somewhere else. When you become a world-class violinist after thousands of hours of practice you don't become clumsy or less functional in other areas."

Not only was I worried about the cost of possible lost brain functionality, I also worried that the time I was expending studying the brain—as fascinating as it was—was coming at too steep a price. I found myself neglecting my work and other responsibilities. Reading articles about the brain didn't get the bills paid, and it didn't keep our customers happy at work. Learning about neuroscience wasn't making me dumber about cars, but it was distracting me from my job, and I could already see how success in one arena might lead to failure in another.

"But that's not getting dumber in the old area. That's changing your priorities," Alvaro told me. My friends agreed that those two things were unrelated and that choosing between them was a simple act of willpower. Unfortunately, it wasn't for me. The TMS effects were far too powerful to ignore. They filled my head with new thoughts and turned my mind in new directions. That's how I felt; I wasn't getting dumber, but I was surely becoming different.

At fifty years of age, I stood atop a pyramid of my collective life experience. A large part of its foundation was the car company I had built. Could I remain in place, and maybe climb a little higher, or was I going to leap sideways into space, fall to the bottom, and be forced to start the climb anew or, worse, build a new pyramid? "These new feelings are a big deal for you, but they don't take away any of your former abilities. Not wanting to do something is different from not being able to do it," Alvaro told me, and I realized it was true. So far, the TMS had been a big distraction, but I could not think of anything good it had taken away. And as soon as I had that thought, I felt guilty, as I remembered my unraveling marriage and how I now lived my life on the run from depression. One reason I was spending so much time in Boston with the scientists was that I couldn't bear to be at home with my newly heightened awareness of what I now saw as my wife's eternal sadness.

I was beginning to realize that my marriage was not as solid as I'd imagined it to be. *Maybe all marriages are like that,* I told myself, but I knew there was a fundamental problem. *Is TMS opening my eyes to a weakness that was always there?* Martha sure didn't think so—she thought the TMS was the problem, and she was scared, angry, and unsure where to turn. I was embarrassed to discuss my marital problems with the scientists, and it never occurred to me that a marriage counselor might be able to help. Instead I ran from our problems and wrapped myself in a cloak of hope that my newfound emotional intelligence would save everything. Ironically, it was turning out to have the opposite effect.

Without any idea of the disastrous effects TMS was having on my marriage, Alvaro remained ever the optimist. He believed the brain naturally rewired itself for better performance, and he thought it did that every time we got really good at any new task. That mental tune-up was part of what researchers call brain plasticity. "So did TMS just dramatically speed up the process of improving performance?" I asked.

"Not exactly," he answered. "TMS probably didn't build new pathways.

Pathways are physical things, like roots beneath a tree, and growing them takes time. That's days or weeks—more than overnight. It's more likely that the TMS energy opened up paths that were already there. Whatever TMS turned on in you must have been there all along for it to come to life as it did."

Alvaro then described for me what he called his theory of "metamodal" organization of the brain,* which argues that brain regions should be defined by what they do and not by what they act upon. The visual cortex has traditionally been called by that name because it processes image data in people who see. If that's all it was, he argued, it should be dormant in a person who is blind. But it's not. Instead, a blind person's visual cortex processes information from the ears, and it extracts data from that sound that gives some blind people an ability to almost see in the dark. It can also adapt to process touch signals from the hands, allowing some blind people to read braille as fast I can read printed words.

When the brain's so-called visual processing center is fed sound instead of image data, it processes the sound stream just as sighted people analyze visual images. We use our eyes to decide which objects are in front and which are in back. In a blind person, that same brain area may tell that person where things are by their sound, or even by the way they reflect sounds—like the echolocation ability of a bat.

In a later conversation, Michael Wilcox shared the experience of a friend who'd experienced that very thing. He told me about his buddy Mark, who was blind, though not from birth, and who went through training to learn how to use echolocation. His guide took away his cane and taught him to walk down the street without bumping into lampposts. He learned to do that by listening extremely intently and paying close attention to where every little sound came from. That allowed him to make a sound-based picture of the world around him, just as the rest of us do with visual signals. He said it took an extraordinary amount of effort, but it also made him realize the possibilities. Mark still uses a cane, but he

---

* "The Metamodal Organization of the Brain," *Progress in Brain Research* 134 (2001): 427–45.

also has an amazing map of the world in his head, and if he's been some-place he can find his way back there. "I was once giving him a ride and my GPS shut itself off for some reason. 'Don't worry!' he said. 'Just take the next exit and then go left at the stop sign.' He guided me to our destina-tion more efficiently than my GPS ever could have."

Alvaro had established something similar in a study that was published a few years before the TMS autism work. He'd set up an experiment in which he blinded people for a week with bandages and watched the activ-ity in their visual cortexes. As expected, those areas became inactive, but they reawakened within a few days as the former visual processor of the brain began evaluating sound data instead. He believed that showed pre-existing alternate paths coming into use and then being quickly opti-mized to process a different kind of data.

What might have happened if he had carried the experiment on for a year instead of a week? Maybe what neurologists called the visual cortex would have become the auditory cortex in those people, and it might pro-cess touch for others. If they'd worn the blindfolds for a year, would their visual processing have come right back when they removed them, as it did after a week?

"Maybe not," Alvaro agreed. "The new paths might have taken over by then. It might have been harder for those people to switch back to seeing."

Several months earlier I had suggested to Alvaro that perhaps the area that recognizes emotions in people was recognizing traits of machinery for me. He had dismissed that notion at the time, but when I asked him again late that summer, he said, "Yes, you suggested that before and I find it very intriguing indeed."

Our current understanding that every brain area is active seemed to support my hypothesis. If I lack a nonverbal conversation data stream because I'm autistic—or if my stream is very weak—the parts of the brain that should have analyzed that data are going to do something else with their computing power. Perhaps my brain chose to analyze machines. That might explain why I pick up on mechanical things others never see, and why I'm blind to the social signals that are obvious to most people. Following that line of thought, if TMS helps me develop an ability to read people, and that ability uses the part of my brain that was reading machines, I might well be trading one ability for another.

A new postdoc in Alvaro's lab helped shed some light on whether that might actually be happening. Ilaria Minio-Paluello, an Italian neuroscientist, had recently joined Alvaro's group. She too had an interest in autism, and she asked me to take part in some studies of her own. One of them proved particularly significant in light of the metamodal idea.

"Have you ever heard of prosopagnosia?" she asked me one day.

When I said no, she proceeded to explain that it was the medical term for face blindness—the inability to recognize faces and people. I'd certainly heard of that, but what did it have to do with me? As it happened, she had a test.

She sat me down in front of a computer and showed me one face after another. There were various exercises, which went something like this: "Take a look at this face. Remember it." I found myself staring at an ordinary-looking face with a neutral expression. Then the faces began changing. One after another, new faces began to appear, with varying expressions. "Now tell me when you see that first face again."

I didn't have a clue. It was very frustrating.

Other times, five faces appeared in a lineup. "Remember that first face I showed you?" she would say. "Which of these is the face you were asked to remember?" It didn't seem to matter if the faces in the pictures had similar or different expressions—I still had no idea. After an hour of frustrating failure, I had a new diagnosis: prosopagnosia. *How could I have gotten through fifty years of life and not known that I am face blind?* I wondered.

"Many of the autistic people I study have face blindness," she told me. "It's one of those conditions that often go unrecognized, so we don't know how common it is. We think it may affect one to two percent of the population as a whole, but in my lab work it's way more common among people on the autism spectrum."

After getting over the shock of her news, I considered what her words meant. I realized that I'd always had a hard time recognizing people. After talking with Ilaria, I figured out that I recognize people by context, which meant I was lost when I saw someone in a strange setting. As a joke, I suggested that she put my retired uncle Bob in a police uniform or behind a counter at Walmart and see if I knew who he was. But she took me seriously, and in fact when she rendered a photo of my son as a line drawing I could not pick him out of a lineup of similar faces.

Shocked as I was, Ilaria reassured me that I'd been adapting automatically all my life so that this particular disability had not held me back.

First I was sad, thinking this was yet another way I was less than other people. But after some thought I realized there was another side to it: I recognized cars and other mechanical things far more accurately than the average person. I often found myself saying, when a car drove into my yard, "That's Bob Parker." But I was recognizing the car—not the driver. If I saw Bob in a store, I'd have no idea who he was, which made for considerable embarrassment if he happened to recognize me and strike up a conversation.

I can't begin to count the number of times that has happened. "Hi, John," someone will say, and I turn and say hi in return. I've learned to be polite, even when I have no idea who's talking to me. The person might go on for a minute or more, but at some point, he or she will say something like "You don't recognize me, do you?" If I'm lucky, it will be said in amusement and not annoyance.

Experiences like those are embarrassing, but I never thought of them as illustrating a particular disability. I just thought of that inability to recognize people as one of my quirks. Now that it was officially pathologized I speculated about whether it might have a different meaning. Maybe that part of my brain was repurposed too.

Ilaria's tests confirmed that my ability to recognize human faces was seriously impaired. But that "inability" was only one side of a more complex story. If Mr. Parker had arrived in a black Mercedes sedan and parked it among ten other black Mercedes sedans in our lot, I would be able to pick out his car without fail when I was ready to work on it. To me, each car was an individual. To others, all cars look the same.

Was that more evidence of Alvaro's metamodal brain in action? Was I using the face recognition area to identify specific machines, to my great benefit? For an automotive service manager, the ability to recognize particular cars, and parts of cars, conferred a great advantage. As a social human being I was disadvantaged by my weak ability to recognize people, but from my working technologist perspective the disability it produced was less than the benefit I got from machine recognition.

It would make sense that a human brain would be configured to recognize more than just people with great precision, I reasoned. Art experts learn to distinguish subtle differences in the works of the masters—things that are totally invisible to you or me.

As much as I wanted to believe Alvaro's notion that we could tune our brains to make them more efficient, I could not help but think that whole swaths of my reasoning power were simply deployed to different purposes than they would be in a more ordinary sort of person. And as badly as I wanted to see into people again, I had to face the truth. Not being able to read people had made me sad, but it hadn't made me a failure. The only consequence I could put my finger on was occasional embarrassment. If seeing into a machine was my offsetting gift—via brain repurposing—it had given me an extraordinary advantage in technical work. Recognizing people, in contrast, would only make me ordinary.

It was a real conundrum. I hated the idea of losing my special abilities, but I also felt alone so much of the time. And TMS had relieved that loneliness, at least temporarily. I had a return visit to the lab scheduled for more TMS in the not-too-distant future. But I was torn now that I'd come to believe that changing my brain might well be a zero-sum game.

When I went back to the TMS lab for the last round of stimulations in their initial study, I was ready for anything. In earlier sessions I'd seen my emotions amped up and my senses fine-tuned. What would come next? The answer, to my dismay, was . . . nothing. The first round of TMS study concluded quietly, without a bang.

As much as I'd hoped for further positive transformation, I realized that wasn't going to happen every time. The researchers had told us that at the start. When I asked Lindsay what came next, she said, "We have to evaluate all these results and consider what to do now." For her there was work. For me and the other volunteers, there was waiting, or a return to our pre-TMS lives.

Interestingly, the younger volunteers seemed to just move on after participating in the experiments. When I met a few of them after the study ended they acted as if TMS had been no big deal. It was the older participants—Michael, Kim, and I—who had experienced the most dramatic changes and wanted more. All of us were eager for the next phase, and we wondered why the younger participants didn't share our excitement. I looked at my son and wondered the same thing.

As summer wound down we stayed in touch with the scientists, and with one another, as we waited to see what they'd come up with next. And we waited for an explanation of what we'd already experienced. "You'll be the first to know when we've analyzed the results," Lindsay had told us. Meanwhile the scientists answered our questions as best they could and we waited to see what would unfold.

The Shimmer of Music

NOW THAT THE STUDY had ended, Alvaro was free to discuss the areas they had targeted. Finally he could answer all my burning questions, one of which was "Did you try this on yourself?" To my surprise, Alvaro told me he had tried the stimulations from our study. "They had no effect on me," he said, "but I'm not autistic. When we do the depression stimulation on people who are not depressed, it sometimes has the opposite effect. Instead of making them happier, it makes them more upset."

That sounded strange, but Lindsay explained. "Think of your brain's function as being on a bell curve," she said. "The top of the curve is what we'd call the optimal area. That's where excitability is kept in check and everything is balanced—not too much and not too little. If someone is disabled, and an area is not functioning optimally, TMS could push him or her to the top of the curve. But if we did the same TMS on someone whose function was already optimal, we'd push him or her off the balanced middle and into less functional territory."

That was a good explanation, as far as it went. But what if a brain area had several different functions, each with its own bell curve? And what if some bell curves were optimal already but others were not? That's when things get complicated.

In neuroscience, conventional wisdom used to say that each brain area performs a predefined function. But my own experiences, and the writings of Alvaro and other contemporary scientists, suggest that this viewpoint is evolving. Suppose that I had always used my "emotion recognizer" to see into machines rather than humans. If that was true then it's no wonder that stimulating this area would affect me very differently than it would affect a person who uses that area in the more typical way. And as Alvaro had shown me, our understanding of what different brain areas do is still very incomplete.

As Alvaro's metamodal theory suggested, even when we associate a given brain area with one task, researchers may discover it's also a key to some totally different function. Computer scientists call that concept "distributed processing." That's what Alvaro believed was happening in our minds. If cognitive tasks are processed all over the brain—instead of in task-specific areas—the job of understanding and mapping function becomes exponentially more complex.

Being a writer, I had made notes the night that I "saw" music. On that night I had made a point of writing that the visions I saw were not actual memories of things I'd experienced in the past but rather constructs my mind had created around the music I was hearing at the moment. I was hearing music through the stereo and matching its tones to musical instruments and scenes that played in my head in time to the melody.

It happened so fast and so smoothly that I felt as if I were at a movie, or at an actual show. When I heard an organ, I thought, *Hammond B3*. As quickly as I thought that, I saw the B3—an instrument I've known half my life, even though I cannot play it—and I watched it play the melodies I was hearing. When I heard the singers' voices, I saw them, their faces indistinct yet real-looking, as they stepped up to the mics to sing their parts. As the music changed, the movie in my head followed.

I thought back to when I was engineering the systems that delivered that music and how I saw the music then. Back in those days, what I loved most was designing and building the pieces of gear that made a show possible. I felt great joy seeing them work. It didn't matter what kind of song rang from my speakers; my happiness came from delivering the show successfully. The audience might have been there to have fun, get high, or get laid, but I was there to work. For me, live music was a serious business. Instead of feeling emotion when I heard a song during a show, I felt the subtle signals of my equipment in a way the audience never recognized. If one of my speakers began breaking into distortion, I felt it like fingernails on a chalkboard, but I never had a hint that anyone in the audience had a clue. Like children, my amplifiers made me proud when they did their jobs well, and I cringed when they failed. I actually hurt when they broke down.

In those days, my mind was like an oscilloscope screen, one that flashed beautiful smooth waves when everything was right and hard angular patterns whenever the sound system overloaded or went into clipping distortion. It took thousands of hours to do it, but I taught my brain to match patterns of waves with layers of sound, so I could see as well as hear the structure of the songs. That visual component helped me to envision and then create the sound effects people would love. With practice, I learned to visualize the waveform as a musical instrument that played in my head. Then I imagined how a circuit I was conceiving might change it. I'd mentally pass the wave through the circuit and predict what the result would look and sound like. When I had a circuit pattern that felt good, I'd build it in real life and then test it. Sometimes I would be wrong about what it would do, but the more I did it, the more often I was right, and it was satisfying to know I could successfully imagine how electronic circuits could bend and shape complex waves of music to make new and better sounds.

At the time, I thought that was just being an audio engineer, but now I know most engineers don't come up with designs in that manner, because their brains don't work that way. For most people, there is no path in the brain that connects circuit diagrams seen in a drawing to sound effects as they are heard. A typical person can imagine and sketch out a circuit, but doing so won't cause him to hear its sound in his head.

Yet that's what happened to me. Neuroscientists call that ability synesthesia, and they say it may be a result of brain areas being cross-connected. However it works, it seems that my math, sound, and visual processors may be a little more interconnected than the average person's. Some synesthetes can taste sound or see numbers as particular colors or shapes. Daniel Tammet, another autistic adult, wrote a book called *Born on a Blue Day* about his ability to see numbers and words as shapes and colors. Neuroscientists believe that as many as 4 percent of humans may have this ability. But I was the only one in this particular study—as far as we knew.

It has always bothered me that many people, doctors included, tend to view anything that deviates from the typical as being abnormal or broken. The common medical perception of synesthesia illustrates this perfectly. Doctors don't generally say, "This is an incredible gift; how can we give it to other people?" Instead, they say, "What went wrong in this poor fellow's head, and how can we fix him?" It made me sad sometimes to realize that the things that made me special were—to many observers—just broken circuits. The TMS was beginning to open my mind to lots of ideas, one being a deeper appreciation of how my brain is wired and what that could mean.

The initial TMS study had ended that first summer, but already the researchers were planning more studies, building on what this one had shown them. "It would be great if you participated in those too," Lindsay told me. Although I had given a lot of thought to the possible trade-offs that might come with TMS, after much debate, I decided there was only one answer. Shamans and scientists both choose a life dedicated to the pursuit of knowledge, and so would I. And the path I chose in this case would lead, I hoped, to deeper self-knowledge, and maybe even a glimpse of the unknown territory inside myself. I resolved to stay with the experiments as long as the studies lasted. Interestingly, when I talked to them later that summer, I realized that Michael, Kim, and my son had all come to the same decision. Each of us felt ourselves affected by TMS, and we wanted to stay the course to see where the scientists could take us.

When I thought of my synesthesia in light of the TMS study, I wondered whether the TMS might have energized connections all over my brain—not just under the coil—and if some of those connections had in turn stimulated the process of synesthesia. Alvaro had almost suggested as much when he told me that TMS wasn't making new pathways, that "it turns on what's already there." And he'd told me that one neuron could have connections to a thousand others, so it was easy to imagine everything connected to everything else, at some level. If I have more or different paths than the average person, it would be no surprise that I'd have broader and stranger experiences in response to TMS.

Most people don't have so many cross-connections in their brains, so they don't see images when they hear a song, as I do. It's even impossible to know if the estimated 4 percent of the population thought to be touched by synesthesia is accurate; the actual number could be much higher because some synesthetes view their way of being as natural, and they'd have no reason to report themselves to a psychiatrist. "There might be a thousand people with minds like yours, working on farms in India, and we'll never know," Alvaro told me. Maybe everyone has a touch of it, and it's only the heavily cross-connected who are truly uncommon.

When I mentioned that conversation to Lindsay, she told me her former professor felt the same way. "Rama would say that's why we call cheese sharp, or a color loud."

Think about how people describe music. It's brilliant, shimmering, and dark as the night. It's light as a feather. The more artistic the person, the more her adjectives are those of sight, touch, or smell. I used to think that was just speech, but now I'm not so sure. When I turn on the stereo and close my eyes as the singers sing, I do see the shimmer. As the music plays, the melody transports me into a world of other senses. Can I possibly be the only one affected this way? Maybe it's a question we are just too polite to ask. Perhaps all of us see music, and it's just a matter of degree.

Aftermath

WHEN MARTHA SAID that I wouldn't need her anymore after that April 28 TMS session, her reaction came as a shock, and it took me a while to process. My fondness for her was not changed by TMS, but the dynamic of our relationship was definitely altered. They say that people either grow together or apart, and I think that is meant to apply to the normal gradual pace of change in a relationship. What happens when the emotional balance between two people tips overnight?

I had always told Martha that she was my guide to the world of other people. She was my interpreter for what others were feeling when I was at my most oblivious. Sometimes this happened in real time, like when she'd say, "This is boring. Let's talk about something else!" That would be my cue to stop talking and let someone else change the subject. Other times she would explain things afterward, telling me that Sam had been very interested in what I was saying, or that Doug didn't seem to like me so much. But now I was picking up nonverbal cues on my own. I was responding better to strangers—not by practice but because TMS had precipitated a sudden change—and as a result I was talking to more people, and doing so successfully.

Martha hadn't shown much interest in what I did "out east" at Beth Israel Hospital. Looking back, I think she was probably scared and insecure. Improved as I thought my perceptions were, I didn't pick up on that. All I saw was an increasing distance between us and I could not figure out what to do. So I did nothing, which was, in retrospect, my worst possible choice.

Her lack of interest may have been based on fear, but combined with my sense of panic about what was happening between us it became a kind of self-perpetuating cycle as I began confiding in others when she didn't want to talk. That too was a bad choice, but I couldn't see it at the time. I was so taken with my newfound ability to relate to strangers that I could not imagine it might have a downside. Now I see that we only have so much time to spend, and that when we take hours away from family relationships to cultivate new acquaintances, family can suffer.

I wished there was help for the relentless cloud of Martha's depression, but neither of us knew where to turn. People sometimes described me as "suffering from autism," and I'd feel offended because suffering was not the right word for how I lived my life. But Martha's situation was different. She did suffer from depression and wished very much to be cured. Unfortunately none of the medicines she tried did more than partly lift the veil that hung over her so much of the time.

I'd been able to live with her depression in my pre-TMS oblivious state. She was always good to me, and we worked well together, and that was enough. We built a successful business together and what I thought was a successful marriage—until TMS made her depression painfully apparent to me.

It's not that I did TMS and a light came on that said, "She's depressed!" It was actually something far worse. The TMS opened me up to receiving the feelings of others, and when I was at home with her, I felt like I was being crushed beneath the weight of her depression. I wondered if the depression was in me, but when I went elsewhere, the weight lifted. Whenever I was with her I felt it was pulling me down.

I was also having new feelings about myself. Before TMS, I seldom gave my own actions a second thought. Now I found myself looking back over things I said and did, and I felt tremendous shame at some of what I'd done. For example, Martha used to be so concerned about her weight and being trim that I nicknamed her Chub. Cubby and I thought it was funny, and she seemed to take it in stride. But now I saw my behavior in a whole new light. It's like the post-TMS me looked at the pre-TMS me and saw a bully. I felt so ashamed it was hard to face her. I now felt as if I was taking advantage of her weakness because she was depressed. Never before had I looked in the mirror and imagined myself as nasty. Now, I wondered if TMS was helping me see myself as I really am—at least a part of me—or if it made me imagine a monster that was not really there. I resolved to watch myself closely in the future and be careful not to be mean.

Martha and I are partners in the car repair company and went to work together most days. But throughout our marriage we had gone through periods in which she didn't get out of bed in the morning. On those days, I would get up and go to work alone, telling our staff that she was working from home, and return to find that she'd stayed inside all day. The staff never thought anything of it, and neither did I—until TMS opened my emotions. I'd always wished she wasn't sad, of course, but her feelings didn't drag me down. They just meant she wouldn't go with me that day. In my autistic-oblivious way I just took the world and other people as I found them.

Now I realize that autism had made us compatible. On the days that she was too sad for most people, I didn't notice. "Come on, get going," I would say to her, and sometimes it worked. But whether it worked or not, I went about my day whatever mood she was in.

After TMS, our whole dynamic was changed. Just as I seemed to be soaking up the emotion in a newspaper story, the same thing started to happen at home. When Martha had her down days, I was no longer able to jump out of bed and go to work. As soon as I got up I'd feel panic over her sadness and then feel sad myself and wonder what I was going to do. Like her, I began to feel that I couldn't go to work, and I was starting to see my life as a failure. I found myself not wanting to be at home, and that was very troubling, especially since we'd just built a new home together. My first instinct wasn't to leave. Marriage handbooks claim that women listen to problems while guys try to solve them. I wanted to solve this problem, but I couldn't. I found myself wondering how I could possibly go on. I kept absorbing the sadness that weighed her down. *I'll get used to it,* I told myself, *and things will go back to the way they were before.* But that didn't happen. I began to have thoughts like *Maybe we'd both be better off dead.* Then I'd get out of the house, and a few hours later, everything would be okay.

I tried puzzling it out. *Maybe the changes from TMS are like the changes when you quit drugs or drinking.* I'd never kicked a drug or drinking habit myself, but there was plenty of that in my family and I'd certainly seen how relationships could change very quickly when habits changed, drinking buddies evaporated, and there were no new friends to take their place. Ex-drinkers could find a supportive community in AA. What sort of supports might I need, and where was I going to find them?

Martha was not the only person I saw differently. Thanks to my new-found power of perception, so many of the people around me seemed changed that I began to feel lost. First there was the sadness at home. Then I began to get the sense that people I'd come to know in the course of my business and my daily life were different. Logically, I knew they weren't (I was the one who had changed), and there were friends like Bob and Dave who felt constant, but my perception was altered and there was no going back.

Richard,* for example, had been one of my best friends. We first met when he was a customer—he'd taken a special midlife interest in our kind of motor vehicles. He'd sit in the waiting room and talk to me while his car was in for service, and we got to know each other fairly well over the years. Our friendship expanded into other areas and we began getting together regularly outside of work and going to dinner with our wives.

Richard worked with teenagers, and he had a way of understanding people that I found fascinating. Over the course of our friendship he had often helped talk me through social interactions with others. We got to

---

* Name changed to protect his identity.

know each other shortly after I learned about my autism, at age forty, when I was feeling especially in need of counsel. It seemed like relationships—no matter how casual—were always full of mystery and missteps.

Our friendship came to symbolize the social success that my newfound knowledge of autism had given me, and we'd known each other for some time before I met Alvaro and Lindsay. I was excited to tell him about our initial conversations and the promise of TMS, but to my surprise, he dismissed them with a harshness that took me aback.

"Stay clear of people like that," he told me. "I know the kind. Neuroscientists have no heart," he said. "They just want to study you and they don't care what happens in the end. You'll end up a wreck, but they will get to publish their paper, and that's all they care about." My first thought was that he was concerned for me. But the way he vilified the neurologists—even though he'd never met them—and his need to bash them every chance he got seemed strange.

He was not the only skeptic among my friends and family, but he was the one with the most carefully thought through objections. Everyone else just said, "Zap your brain? You're crazy!"

When I asked Richard why he was so dubious, he said he was just looking out for me, and sometimes that included telling me things I didn't want to hear. A true friend might do that. But I'd also known instances where people who weren't friends at all told me things "for my own good," when in truth they wanted to take advantage of me. So my antennae went up a bit even though I could not zero in on what was bothering me.

His refrain—"They are just using you and don't care how you feel"—did not seem right. I tried to defend what they were doing with my budding knowledge, but my defense became a provocation. "You don't know what you're talking about," he told me. "You're well qualified to talk about your life as an autistic person. But when you start talking about how your brain is working, or giving people advice . . . you're out of your depth. That's a job for experts and you are not an expert in the brain or how it works."

I wasn't sure what to say. I was very confident of my understanding of how TMS delivered energy into the brain. Beyond that there was much that I, and the scientists, didn't understand. But was I truly out of my depth in expressing dreams for what a technology like TMS could accomplish? Richard's criticism left me at a loss. I knew my ability to read people was weak, and I had always trusted his opinions.

Since the TMS, I had been noticing new things about my interactions with others. I found myself sensing what they felt or what they wanted out of an exchange. Or rather, I believed that I was. But since I had lived life up to this point misjudging and being emotionally oblivious, I could not be sure that the new signals I picked up were right.

I monitored these changes closely, as much to evaluate myself as those around me. What I saw and felt from Richard was very troubling. He would belittle me in front of people, seemingly for his own amusement. Often, when I met him with a group of other people, he'd greet them and then me with some remark like "And you too, John." I'd never made anything of that before, but I now realized he was singling me out, separating me, and that kind of separation had been a pain I'd felt all my life. Why did he do it, especially with what I now recognized as a little smirk?

He'd tell stories right in front of me about how I'd overcharged him for fixing his car, or fixed his car wrong, but then he'd smile and say, "We're best buddies anyway." He'd look at me as if it were a joke, but there was nothing funny from my perspective. If that was how he felt, I reasoned, he had no business having us fix his car. The realization that he'd been saying that kind of stuff for a long time—years—made me feel even worse. I felt like a boy again, when the older kids would say, "Get out of here, retard! We don't want you around!" Then a few minutes later they'd say, "We're just kidding," and I would smile my biggest smile because I wanted so very much to be liked.

Things came to a head on New Year's Eve, when Richard joined a few of us for dinner and revelry. Cubby was with us that night, along with his friend Masha. Cubby and Masha had gone to high school together, and she lived a few miles away. Masha was born in Russia, but you'd hardly know her from your average American teenager. Her mom taught Russian at Amherst College and her grandmother had taught there before her. Growing up, my friend Aaron had Masha's grandmother as his Russian teacher, and he used to tell me stories she'd shared with the class about her husband, who was a refugee from Stalin's gulags.

As the evening progressed, Richard began telling jokes about Masha and Russians. The longer it went on, the nastier they became, and the more he drank, the worse his behavior was. Another of my friends actually got into an angry exchange with him. By the end of the night I'd come to a decision. Swallowing my anger and hurt, I went home quietly, as did Cubby and Masha. But I would never speak to Richard again.

A month later, Richard sent me a long, rambling email. In it, he said he had been drinking a lot lately, and that when he drank, he said things he should not say. He reaffirmed his fondness for me and said he missed our friendship. But it was too late. With the light of TMS illuminating the history of our relationship, I saw a pattern. There had been too many little digs over too many years. Could I forgive him? Sure. But that didn't leave a basis for restarting the relationship. Humiliating me for his own amusement was bad enough, but realizing that he had done it when I'd been too oblivious to understand what was happening was even worse.

Losing friends hurts, and even writing the story years later is painful. TMS took away my emotional innocence, and I'll always be sad for its loss. But part of the cost of getting smarter emotionally was seeing people as they actually were, and not as I imagined them to be.

I didn't feel any different about my son, Cubby, but I did see his place in the world in a different light. When I watched him interact with other people I now saw his autism clearly—when he failed to respond to a social cue that I recognized, or when he went on and on about something and his listener was bored to distraction. Watching him made me realize that I must have been similarly oblivious to my own behavior—perhaps even worse. And it made me wonder if I was still doing those things today.

As I pondered what was happening, the summer of 2008 arrived, and the initial TMS-autism study came to an end. Almost imperceptibly, the effects of the sessions faded. There was a gentle leaving as I stopped noticing emotions in others. It's funny—I had been oblivious to other people's emotions my whole life, but now, after a brief taste of this new sensitivity, I felt a huge sense of loss as seeing feelings slipped away. And every now and then I felt a flash of fear too. If my emotional intelligence had been ratcheted up by TMS, might the rebound take it even lower than it had been at the start? "I doubt that would happen," Alvaro said, but his reassurance was limited because none of us knew for certain. The effect they'd seen in me was, in his words, "hoped-for but unexpected," and it was impossible to predict what would come next. I remembered what I used to tell my son when he said he didn't believe in monsters. "The kids that know the truth are all gone," I said. "Eaten." He said he didn't believe me, but he wasn't completely sure. Should I be worried now?

I decided the best thing to do was to fight the loss. I would look deeply at every person I encountered and will myself to read and feel his or her emotional state. I was of course using logic to force emotion, and it didn't work. The feeling of emotional ESP evaporated despite my best fight to keep it. But as the summer wore on, something different, and more subtle, started to happen. An improved ability to read and engage other people was again building in me, slowly but surely. I was left with the sense of a new primal connection to my fellow humans, a feeling that was unfamiliar to me, yet also perfectly natural. There was a new current of emotion in stories I read, and movies and television shows became such a roller coaster for me that I finally stopped watching. The people at work said I seemed more expressive. They'd say, "Customers have noticed!"

I asked Alvaro what it might mean that I was still changing, though any energy they'd put into my brain was surely long dissipated. "The TMS might have opened some doors for you, and now your mind is passing through them. The TMS energy is gone, but perhaps you used the paths through the doorways enough while it was there to keep them a little bit open. Let's see what happens as time passes."

"Has anyone in your studies had permanent change?" I asked him.

"The closest would be the depression patients," he told me. "Some of them can go several months before the stimulation needs to be repeated. And there's some sign that the effects may last longer the more we do them. It's too early to tell. Also, at some point, the effect of TMS becomes indistinguishable from the brain's own process of plastic change. You've already shown a remarkable ability to change, and this may be more of that."

My newfound ability to look into people's eyes without discomfort lingered much longer, perhaps six months, and left me permanently altered, though today this ability is much more limited than what I had under the direct influence of TMS. Today I can do it well enough that no one describes me as looking at the floor or ignoring them in conversation. Seven years later I feel I can call that a permanent change.

The other thing that's permanently changed is my ability to converse with strangers. When I meet someone new I'm able to follow and mirror her conversation far better than I ever could before. When I was a boy, kids would approach me and say things like "Look at my new dump truck" as they held out a shiny toy. Instead of admiring it and saying, "That's a nice truck," I'd answer with something inappropriate like "I like helicopters" or "I like elephants." Needless to say, those early social exchanges mostly ended in failure.

Maripat Jordan was one of the first to notice my transformation. Maripat was a well-established figure in the Springfield media community when we first met twenty-some years ago. She published a regional business magazine and had headed up sales for several big TV stations. Most business owners in my area knew her, or knew who she was, because they bought advertising. But I didn't because I was still new in business and I'd never advertised.

One day, Maripat drove up to Robison Service in the hope of changing that situation. I walked out into the yard as her car pulled in—a minivan. As a specialist in luxury cars, we didn't get many of those in our parking lot. As there was no telling who or what might be inside, I approached cautiously. Out stepped a petite, short-haired woman.

"What do you want?" I asked. Neat and trim, wearing dark sunglasses, she did not look very dangerous. Consequently, she merited a pleasant greeting, which I believed I had just delivered. I also thought she was kind of cute, but you don't say things like that to strangers.

"I knew he was the owner," she'd tell friends later. "No employee would have ever been that rude. He'd have been fired."

Later on, when we had gotten to know each other, she revealed that she was somewhat taken aback when she met me. I asked what had been wrong with my introduction. I had, after all, greeted a thousand visitors before her in exactly the same way. Was there something better I could have said?

"Yes," she told me. "What about 'Hi, I'm John Robison. How can I help you?'"

Her suggestion left me momentarily speechless. Asking how I might be of help to her rather than demanding to know what she wanted had never occurred to me. What a great idea! I've tried to do it her way ever since, whenever strangers appear in the yard, with generally beneficial effect. I've now learned that good things can emerge from the most unprepossessing of motor vehicles, so I try to act accordingly.

After that first encounter, we ran into each other off and on through the years, more so after she bought a secondhand Audi from our company. I thought it was a rugged car but she came to see it differently—as constantly in need of service. Still, I always liked her and was happy to see her when she came around to get it repaired. We generally got along, but as she would tell me later, "You always acted kind of different." Then I saw her by chance after the TMS experiments, and she remarked on the striking change.

I'd walked into my local Panera and was standing in line when I spied Maripat in a booth on the other side of the low wall. "Hi there," I said, and then I noticed she was sitting with a group of people. I looked at them and said, "I'm John Robison, from Robison Service on Page Boulevard." Each of the people introduced himself or herself to me, and I spoke to them for a moment before returning to my place in the line. The whole encounter had seemed perfectly natural and smooth-flowing to me, but in retrospect I realized how remarkable the exchange was. I'd never had the ability to engage strangers so easily before. I could always say hi and turn away, but actually shaking hands with strangers and engaging them in conversation was totally out of character for me, at least before TMS.

As Maripat told me later, "I thought to myself, *What happened to John? That was a perfectly ordinary thing to say, but he's never said ordinary things before.*" I was a little embarrassed to hear that, but it was one more piece of confirmation that TMS was making me more successful at engaging strangers.

The only snag in the Panera experience was that when I returned to the line after greeting Maripat and her friends, a few patrons thought I had given up my place, and I had to set them straight. Luckily she didn't see that part.

I got my sandwich and iced tea and retreated to a quiet corner table where I was out of sight and could eat and read in peace. The TMS sure had changed me, I thought, but now I didn't know what to do next. Should I have asked if I could join them? Or should I have gotten my order to go and left? If I'd learned manners as a kid, I might have had some idea, and I wished I'd paid more attention to my grandmother. I read my Emily Post and tried to catch up.

The next time Michael Wilcox and I had lunch, I told him what had happened. "I haven't seen any changes like that in myself," he told me. But when I looked at him, his directness and connection to me was obviously better than it had been before TMS. When I pointed that out to him he said, "Yes, I do feel sharper. You're right about that."

Then he said, "You know, I never knew you felt so broken before we started this study. Maybe you went into it looking for a lot more than I did, and you found it. I just approached it as research without expecting anything to happen."

I realized he was correct. "You were able to graduate from school and get a good job," I told him. "I could never do that. No matter how successfully self-employed I was, I always felt inferior because I didn't fit in with regular society." Once again, I was struck by the difference between how I saw myself and how others saw me.

"That's why you write about it," he told me.

But it wasn't just my emotional awareness that was transformed over the summer following the TMS study—my overall state of being improved markedly. I stopped getting anxious and worried about every little thing, as I had all my life. I could even let go of the depression I absorbed from Martha, once I was out of the house and in a different environment. Somehow, the TMS had broken my mind out of its tendency to get stuck in circles of negative thought. When something bad happened before TMS I would often perseverate on it and be upset and worried for days. Afterward, I still perseverated on bad news, but I generally broke out of it by the next day, and that was a big improvement for me.

The range of benefits I attributed to TMS seemed awfully broad, and I wondered if some of that was just wishful thinking. Alvaro wasn't sure, but he did observe that getting happier sometimes seemed to change everything in a depressed patient's life, and what I described was no greater an effect than what some of them reported.

The durability of the changes in my brain shows something else remarkable. It shows the power of the mind to seize on something good and reinforce it on its own. The TMS showed my mind a better way, and my mind built upon that with a process of rewiring that continues to this day.

Nature's Engineers

"SOMETIMES I THINK of autistic people like you as nature's engineers," Alvaro told me. "Thanks to the differences in your brain, you were able to observe things and teach yourself engineering skills without the benefit of a university. You're able to look very deeply into machines and see points of strength and weakness. You don't necessarily know what to do in social situations by instinct, but you can teach yourself by reasoning and practice and it's worked. That's obvious looking at your life. Whatever your social problems may be, that would have been a very powerful gift in a primitive society."

I liked those words: "powerful gift." The notion that autistic people ruled the world in ancient times was appealing, though of course I had no evidence that it was true.

Whenever I was in Boston, I stopped by the lab, and if Alvaro was in we'd sit and talk. One topic that we seemed to return to was whether autism was an evolutionary adaptation or something gone wrong, and Alvaro's opinion on the matter took me by surprise: he told me he wasn't sure. Some of us seem to have unique gifts mixed in with our disabilities, but many of us seem mostly disabled by autism.

I was beginning to see the same thing in my travels. People like me seem disabled because we often can't do schoolwork in the structured manner that today's institutions require. But had I been born in an earlier time, with less regimentation in my schooling and more focus on hands-on learning, I might well have been an academic star instead of a dropout.

I realize that may not be true for everyone on the autism spectrum. My gifts are mostly of a technical nature, but there are others with autism who think they have no special gifts at all and believe that their autism is purely crippling. I don't mean to paint an unrealistically positive picture of what is to most people a disability, but at the same time I'm reminded of how little people thought of my prospects as a teenager and of how marginal my son seemed to the psychologists who tested him at age six. In addition there's the fact that my writing, storytelling, and disability advocacy only emerged in my sixth decade of life!

When I think of that, I realize we cannot know the future, or the potential, of anyone. Maybe I am intellectually gifted, and some people surely are intellectually challenged. There should be a place for all of us in this world, but modern society makes it pretty hard for those of us who are different, no matter how smart the tests say we are.

No one had ever called me a natural engineer before, but it was true. Details in machinery and electronics that were invisible to the average person were obvious to me.

"Look at your understanding of math," Alvaro said. "As you've told me, you can't do anything but basic mathematics with a pencil and paper. Yet you could see musical waveforms in your head, combine them, and correctly imagine the result. You might not have been able to write that as an equation, but you could solve the problem to a close approximation."

Alvaro's words sent me back to the years when I worked as a signal processing engineer. I'd felt such a sense of wonder the first time I saw a sophisticated tool—an FFT spectrum analyzer—a machine that could do analysis like I did in my head and show it on an oscilloscope screen. The computer showed so much more than I could see through imagination alone. And it had great precision. You could put the tool's cursor over a point and read it out to six decimal places. But the one thing it could not do was show me the actual design solutions I wanted. It just showed data and couldn't tell me what to do, so in a problem-solving sense, it was useless.

My own vision wasn't like that. A typical electrical engineer who designs things using formulae from a text would follow a path set out in the book. He'd choose a standard design, calculate the necessary component values, and come up with exact, repeatable numbers for his circuit. I didn't design things that way. Instead, I was more likely to imagine a new circuit topology and then guess at the initial component values. I never saw great numerical precision; rather, I saw approximate values I could fine-tune for the exact result. My process seems more random, but if the goal was originality—like it often is in music—my creative approach gave me a decided advantage. That was the way I'd seen equations in school, where it wasn't so useful. My teachers criticized me for not doing the work in the accepted way, but out in the real world results were what mattered. For me, adjusting a circuit was like tuning a guitar—I could feel when I had it right. Musicians value that ability, and I can still remember how it was first revealed to me.

It was at the Blue Wall at the University of Massachusetts—a place where all the local bands performed and some got their start. The drinking age was twenty-one, and I was fifteen, but the guys at the door were college students and they let me by when I said, "I'm with the band." Some musicians I knew were getting ready to play and I had drifted backstage to watch what they did up close. One of them was tuning guitars as I watched. He plugged the instruments one by one into a tuner, which was a device with a spinning disc that seemed to stand still when you played just the right note. I watched him tune two guitars, plucking the high E string and turning the tuning peg.

"You always tune upward," the musician told me, "by tightening the string." As I watched, he twisted the peg and suddenly the spinning disc stood still. How it happened is a mystery, but that frequency imprinted itself in my mind. I picked up the next guitar myself and tuned it right to the spot by ear alone. Then I tuned the other strings to that one. Correctly.

"That's really cool," he said. "You must have perfect pitch." Then he stepped out onstage and ripped into an Edgar Winter song. As the notes poured from his guitar I marveled at the relationship between them, how they sounded right when they were in tune and subtly wrong when they were off. From that night on, I could tune guitars.

Michael Wilcox told me he experienced something very similar. He used to develop mathematical formulae for financial analysis. His ability to do that was exceptional enough that he had a group of mathematicians working for him on Wall Street back in the 1990s. "Sometimes they would get stuck on a problem," he told me, "and I would just see the answer. Go try this, I'd say, and they'd look at me like I was nuts because it just came out of my head, but most of the time it was right." Michael never worked the answers out on paper; he just saw the paths to solution. Guided by his insights he solved complex problems in financial analysis without ever knowing how he did it. "I used to assume everyone was that way," he told me. Both of us agreed that this gift was instrumental to our success.

Unlike me, Michael had actually gone to college and studied math, but he still solved equations in his own way. And like me, he never realized that was unusual until much later in life. No one knows how many people have abilities like ours. Perfect pitch is fairly common among musicians. Clearly some mathematicians and engineers have the gift of just seeing answers. Researchers are beginning to think that perfect pitch is more common among autistic people, just as autism is more common among musicians than in the general population. And seeing answers to equations is more common too, as evidenced by the many autistic people with unusual math skills.

Neuroscientists can't explain why this is. It may be that we solve problems that interest us by deploying brainpower that typical people use for other things. With a billion neurons per cubic inch, a fist-sized chunk of brain has far more latent computing power than any device ever built by man. Maybe we've just got the gift of harnessing it in a particular way.

"Do you think my ability to visualize musical waves is like the calendar-calculating ability some other autistic people have?" I asked Alvaro.

"I don't know," he answered. "We don't know how that works either, so it's impossible to tell if they are the same or related."

Consider Isaac Newton, widely acclaimed as the inventor of calculus. Several recent articles suggest that he was autistic, as they point to signs of autism in accounts of his behavior. We can't ever know that for sure, but I had a flash of insight into how he perhaps "invented" calculus. Maybe he saw waves in his head, like I do. But he lived in a time when there were no oscilloscopes or electronic aids to visualization. Perhaps he devised the representational system we call calculus as a way of demonstrating what he visualized instinctively to other people, who otherwise would have no idea what he was talking about.

When I suggested that to Alvaro he thought for a moment and said, "Maybe so."

After our conversation I went back and reread the articles about Newton and autism. In one article Simon Baron-Cohen, an autism researcher at Cambridge University, described Newton's behavior in a set of paragraphs that could have been just as easily describing me, except I didn't have a nervous breakdown at fifty. Dr. Baron-Cohen seemed to think Newton was on the spectrum, and after reading his account, I was inclined to agree.

Then I thought about my own history—diagnosed at age forty with a son just recently diagnosed, but who all his life felt different, like me. My father had died a few years earlier, but my stepmother and I both felt sure he'd had traits of autism too. Looking at the rest of my family tree, I realized that the eccentricity displayed by many of my cousins and ancestors points to a thread of autism. And those nonspeaking cousins I grew up with were not "idiots," as my grandfather had dismissed them. They were almost certainly autistic people with more severe impairments than mine. Today we know that autism is considerably more common among engineers, scientists, and musicians. It's also common among lawyers and clergymen, which my family tree has in abundance. The more I researched my history, the more I saw that my own family's thread of neurodiversity might reach all the way back to Newton's era, and maybe even farther. Some of my ancestors were truly exceptional people, but other relatives— even cousins I grew up with—were disabled enough that they spent their lives in parents' basements or attics.

Scientists now believe autistic people have too many connections inside their heads, and that may well be disabling. But it may also be one way some of us are gifted in seeing and creating patterns. Non-autistic brains undergo a kind of pruning process in the first decade of life, where excess connections and unused neurons get removed in what neuroscientists think is a process of optimization. That does not seem to happen in the same way, or to the same degree, with autistic brains.

There are researchers who suggest that these extra connections make us sensitive to sensory overload and others who believe they lead to confusion when brain signals get lost on the multiple pathways. Then there are the autistic people who have extraordinary calculating or other abilities, and they sometimes ask if that may be associated with those so-called excess connections. "Nature's stingy," they say. If you believe autism evolved in us for a purpose, the connections may be there for a reason. At this point, no one knows.

Thirty years earlier my autistic disabilities had prevented me from succeeding in school. And because I'd never graduated from high school, I was not considered qualified for the conventional scientific work I might otherwise be suited for. But I was technically qualified, and I found a home in theater and professional audio. Three hundred years ago a person with my ability to see into machines would have been at an advantage, pure and simple.

I'd been thinking a lot lately about that in relation to the thread of autism that runs in my family. My father seemed less disabled by his differences than I was, but perhaps his differences were just hidden because he grew up before autism awareness, and he became a college professor, a profession in which eccentricity went unremarked. My son grew up in a more diversity-aware world, and he might be more visibly autistic than I am.

When I asked Alvaro if he thought I would have been considered disabled or gifted if I had lived in Newton's day, he smiled. "You probably would have been both, just as you are today. Your social problems would have existed then as now, but your ability to teach yourself might have been an even greater gift in a time where there were no modern teaching tools." Lindsay had gotten her PhD in psychology, and she agreed—my own autism was likely a double-edged sword, now or in an earlier day. Hearing these words, I wondered if the three generations of us were perhaps the same, and those of us who were born earlier simply fit better into a less restrictive world.

Speech

ONCE THE FIRST TMS study was complete, I learned that the frontal lobe regions Shirley and Lindsay had stimulated were associated with Broca's area, a region neuroscientists believe to be the seat of language in humans. The before and after exercises we did measured aspects of speech, and they picked up some changes. My voice became more expressive, with more tonal range and more change in rhythm or prosody. Several other people in the study saw similar benefits. But the big effects from those stimulations—to me—had very little to do with talking. For me, the impact was all about sensing and feeling emotion. That leads me to wonder if the function of Broca's area is not as well defined as neurologists believe. Some scientists are asking the same question after observing near-normal language in people with major damage to their speech centers.

Alvaro and his researchers also thought Broca's area controlled more than just speech. They had chosen to stimulate there because of Broca's relationship to mirror neurons. Broca is one of the key parts of the brain's mirror neuron system, which is tied in turn to social interaction. According to Lindsay, that's probably why I had such strong emotional responses to its stimulation.

Broca's area is a region in the left frontal lobe named for Pierre-Paul Broca, a French physician of the mid-nineteenth century. Dr. Broca examined the brains of patients who died with language impairment from stroke, disease, or injury, and he discovered a strong correlation between lesions or damage to the part of the brain that now bears his name—whose function was previously unknown—and the ability to speak. This marked a major milestone in our understanding of brain organization.

Damage in that region seemed to produce deficits in language, which is now diagnosed as Broca's aphasia, or expressive aphasia. The researchers in the TMS lab had devised some novel theories about how the regions around Broca's area might be interconnected, and specifically how they might be differently connected in autistic people.

A burst of high-frequency TMS pulses applied over Broca's area on the left side would shut down the ability to speak, Shirley told me. This wasn't what they were doing in the autism study—what they proposed was a much subtler tweaking. But I was intrigued by her comment and didn't let it go. "Did you actually try it yourself?" I asked her. It turned out that she had—in fact, quite a few of the researchers, as part of their training to work in the lab, had experienced the speech-suppression TMS. They offered to show me what it felt like.

In an earlier conversation Alvaro had said he believed every scientist or doctor who administered TMS should know how it felt so she could relate to her patients. On a visit to the lab that summer, I got to experience that myself. The speech center stimulation made for a safe yet very powerful demonstration. "It's only temporary," was the last thing I heard before the machine fired a burst of electromagnetic energy into the left side of my brain. With no more than a slight pop, language was gone. The vanishing was so complete, I didn't even know what I'd lost because the entire concept of words and the ability to string thoughts together simply disappeared. One moment there was a dialogue playing in my head, a little voice saying, *I wonder what's going to happen when they do this.* The next moment, the voice was gone. All that remained was feeling. The comfort of the chair, and a sense of familiarity with the people beside me.

I was incapable of a realization like "I can't talk!" since it consists of words. Without words I had ceased to be a creature of coherent logical thought. Instead, I lived in the moment with sound, sight, smell, and feeling. Some say that's how a dog experiences the world, although dogs are much more reliant on their sense of smell, which is far better than ours. And of course they are capable of barking.

Did I truly become like a dog in the blink of an eye? I couldn't have told you in the moment, because all vocalization—speaking or barking—was suppressed. But as I remember it, my ability to understand spatial relations—and complex things—remained intact. I know that because I looked around the lab at drawers and doors and felt a familiarity with their function while my mind remained mute. Opening a door was still a familiar idea to me, even if I lacked the words to describe it. For the brief time my speech was suppressed, I still recognized the physical world around me, and my mind still worked as before, just without language. To me, that shows that you don't have to know the word "camera" to pick one up and take a picture. That said, I would be mighty impressed if Oigy, our Imperial War Pug dog, started using my camera, no matter what kind of stimulation we gave her.

But were we so different, my pug and I? If I'd been without words all my life, like Oigy, how could I have grasped the concept of a puzzle or a camera? Though words aren't necessary to do things like open doors or use complex tools, how could I have learned those skills in the first place if I lacked the words to understand my teachers? That brief insight into life without language gave me a lot to think about.

I've always felt I learn best by doing. And learning by doing could probably take me reasonably far with no words at all. I surely could have learned to use a saw, hammer, and nails and done basic carpentry without language. But I can't imagine how I'd ever learn something as complex as electronic circuit analysis without words to put the technology into perspective.

So there I sat, fully aware but wordless. In my newly altered state I felt many things in rapid succession, but fear was not among them. There was a curiosity that something had happened, though exactly what that was eluded me. As the effect of the TMS wore off, language returned, and with it a sense of wonder that such a profound change could be wrought so easily and so completely.

"What if language had never returned?" I asked, once I could speak again. I thought I'd been silenced for a matter of minutes, but Lindsay assured me it was only around thirty seconds. And I wouldn't have stayed mute, she explained, because the paths of language are very deeply ingrained in my brain. We may disrupt them for a moment, but it would take far more than a brief stimulation to remove them for good.

Later on we would learn that autistic people experience more profound and more lasting change from TMS than non-autistics do. Lindsay researched that issue in several studies that I took part in, the most recent of which ended in 2014. She believes autistic people have more brain plasticity, which causes us to change more, and to stay changed longer, in response to life events. That may also explain why a stimulation that had a brief and minor effect on her might have a much different impact on me, and why I felt it lasted a lot longer.

It was strange that under the influence of that stimulation, it had never crossed my mind to be scared. I'd even felt a sense of peace. Was "scared" just my speech center talking, expressing its own primal fear that the little man in my head might finally be silenced? Would the rest of my mind care?

Loss of language, as I experienced it, was immediately replaced by the expansion of something that was always there—a holistic understanding of the natural world through sight, smell, and sound. That background is there for all of us, but it's usually overpowered by the words flowing through the thing we call consciousness. Some might say the state I was in was akin to one of meditative bliss. I felt one with the world around me, freed from the strictures of logic and spoken language. Yet meditative bliss is a state that's pursued and attained voluntarily. Loss of language is no one's goal.

Try to imagine living your whole life that way. You can't, really. Because living without language would mean living without the thoughts required to form those words.

Still, after pondering the experience and the questions it raised, I ended up with more questions: Does being without speech bring us closer to a natural world or a spiritual one? Might such a revelation—if that is the word for it—become visible after a time of not talking? Is this the reason monks and other spiritual devotees take vows of silence or spend time in silent retreat?

As Lindsay pointed out, all of my philosophical musings were based on one brief experience during which I gave up speech voluntarily. She didn't think there was any comparison between that experiment and being born with major language impairment.

"Where would you be today if you'd never developed the ability to speak or understand language?" I had to concede she was right—and that was a scary prospect to think about. Most everything I've done to achieve and maintain independence has been founded on reading or hearing instructions and taking the words to heart. Without language, I'd have been left with observation and imitation, and I don't think that would have taken me anywhere near as far in comparison.

That was perhaps why she saw her own experience with speech-suppression TMS in such a different light. Whereas I accepted whatever happened, she resisted it. As soon as her speech was suppressed, she tried to speak. She thought she was talking, but no words came out, and she didn't like that realization one bit. "I can't believe you thought it was tranquil," she told me. "I thought it was incredibly powerful and scary."

It was fascinating to consider how differently we had experienced the very same stimulation. The brain is a mysterious thing.

A More Subtle Result

EVEN THOUGH THE study ended, I continued visiting the lab, to talk and learn more. When I look back on that time, it's amazing to observe how the effects of a dozen half-hour TMS sessions had come to dominate my thinking. Before we began I'd believed my intelligence and my senses were largely immutable and unchangeable. The experiments had shown me how shortsighted that view was. After experiencing TMS it was hard not to think that anything was possible, if only the doctors knew where to aim the coil. The researchers were already planning follow-up studies, and I was proud to be able to offer some thoughts about them. I wanted to know if I could recapture the greater insight I had experienced briefly. Alvaro, Lindsay, and I talked a lot about that through the summer of 2008.

I'd been a passive participant in the original research, letting the scientists stimulate whatever area was next on their agenda. Now I took the lead, or tried to, by asking them to "hit that area that made me see emotion again." At first they were reluctant, saying I'd completed that part of the study and they could not do it again. As a research subject, you are expected to accept that. But Alvaro and I discussed the fact that the tests they had devised in the first study had totally failed to capture the greatest effects TMS had had on me. Though research scientists generally can't deviate from the original study protocol that they present to the hospital's ethics board, Alvaro was also a medical doctor who treated patients. That gave him more latitude.

As it happened, the crew in the TMS lab were already contemplating a follow-up study to try to measure some of the effects the first study had missed. Alvaro agreed to speed that up a bit and try it out on me. When doctors use a tool like TMS in ways that are not FDA approved it's called "off-label." That's what my next session would be—the hospital's first off-label use of TMS to treat autism, one whose findings would guide a larger study.

I was excited and hopeful but also a little bit afraid as I returned to the lab on August 12, the day before I turned fifty-one. The possibility of pain or medical catastrophe didn't scare me anymore, but I was still preoccupied with the "zero-sum game" idea, the thought that enhancing my emotional sensitivity could somehow dull my mechanical awareness. In the absence of any proof one way or the other, in the months that had passed that idea had taken firm root in my mind. My new emotional insight seemed like just such a trade-off, given the emotional fragility I'd also had to contend with. I'd quickly learned that it takes practice to handle the strong emotions. In the three and a half months since the study had ended, I was starting to realize that the old maxim "Ignorance is bliss" might well apply to my former autistic blindness, with respect to reading emotions in certain people.

Then there was the question of what would happen when we stimulated an area that we'd targeted already. Every other TMS session had stimulated fresh territory in my brain. We'd never tried hitting one spot repeatedly, though in their depression treatment they'd found that repeated stimulation of an area made the effect last longer.

With that in mind, Lindsay and Shirley planned to reenergize the part of my brain that had precipitated the hallucinations and insight into others. This latest stimulation would be essentially the same as the one before, but there would be an important difference. The first time, my brain may have been primed by the TMS session that had preceded it. Now, it had cooled off for several months and the effects of this experiment would be from the single treatment alone.

When I got to the lab, they immediately set me to work at the computer screen with a battery of tests I'd never seen before. I looked at stick figures, and at photos of whole people and faces on the computer monitor. They asked me to "pick the identical expressions," "recognize the sad face," or "press a button to indicate what this person is feeling." After three months of imagining myself to have more insight into people, their testing put me in my place. Seeing the images flash by on the screen and having no idea what they meant was a big disappointment.

This time, though, I questioned the tests and not myself. *You are different,* my inner voice assured me. I recalled the number of people who had seen me and said, "You've changed! What's up?" Those were people I knew casually; they had no idea I'd done the TMS but they saw a difference in how I related to them. Something must have changed to make those people comment. And it was a good difference—I could sense that myself.

Yet I knew I was failing the tests Shirley and Lindsay had devised for me. *What were the differences between real-life social interactions and the experiments in the lab?* I asked myself. The lab environment was totally artificial. They'd set me before a computer that flashed faces with exaggerated expressions and then expected me to choose a word for what I saw: angry, sad, scared, disgusted, happy, or neutral. We started with a face flashing for 100 milliseconds—a tenth of a second—but that was too fast. It didn't make any sense. "Why one hundred milliseconds?" I asked.

Lindsay conceded it was a somewhat arbitrary choice. "We want to see you choose on instinct, not logic," she said, "so we used a shorter time." Meanwhile, I was unsure if I should blame their timing or myself for my inability to pick the right expressions. I thought to myself, *Maybe after another TMS session I'll get these. . . . Maybe it's just too fast because I'm slow. . . .*

She adjusted the program to leave the faces up a bit longer and I did the tests again with a new set of faces. I still felt I couldn't do it, and it upset me a lot. But was my newfound ability really lost? Had the comments on how I had changed been tapering off these past few months? First I had questioned their test; now I was questioning myself. I'd arrived at the lab excited but now I was scared, and anxious.

Shirley and Lindsay tried to reassure me—"There is no failure on a test like this"—but I'd heard that line before and I knew there were right answers and wrong ones. There's no such thing as a test where the answers don't mean anything.

For a long nasty moment it was as if I were back in high school and the people on the screen were other students making faces and ridiculing me. The one genuine emotion I'd recognized looking at those computer faces was disgust.

In a few cases, I saw a fairly neutral face, and I picked "neutral," but I thought they were brooding and a bit angry too, and most of the others seemed so exaggerated as to feel fake.

Then I thought back to the earlier facial recognition tests I'd done in the lab. Back then when I'd looked at faces on a computer screen, I'd had no idea what I was seeing. This time I was recognizing a single emotion— disgust. And I was recognizing other expressions as "not disgust," even if I wasn't sure what they meant. Maybe I hadn't lost everything I'd gained. One of my therapist friends told me that we often imagine catastrophe when we anticipate what's ahead, and I knew he was right.

We moved on to the TMS room, and I settled myself in the chair. Lindsay positioned the coil on my head and pressed the button. We were once again stimulating my frontal lobe, an area in front of my right ear.

I gazed at the wall—at nothing in particular—as the pulses tapped away at my cranium. The familiar trance-like feelings of TMS returned, and I sat placidly through the stimulation. As soon as the session was done, I turned to the researchers. They had changed places during the session, so Shirley stood behind me, holding the TMS coil. At that moment, Lindsay was seated at the desk, working with the computer. Looking in Lindsay's eyes I said, "Do I look any different?"

"Maybe," she answered. "You're looking at me very intently." Hearing her, I looked away, not wanting to seem rude or intrusive. I realized I'd been gazing into her eyes, and I'd sensed concern and curiosity. Was that what she was feeling? I was embarrassed to ask, as if I'd been peeking into a private space and I had no business remarking on it.

I looked briefly at Shirley, and the feeling was the same. Was this what had happened before? So much had changed, it was impossible for me to tell. I turned to the desk, where their computer test was waiting. As in the original study I would do a series of tests, then receive TMS and do a similar set of tests right after.

As I sat at the computer, looking at faces, I realized something was now different. I no longer felt a personal connection to what I saw. When I'd looked at the faces before TMS I felt that the emotions I saw—principally disgust—were directed at me personally. Now when I saw expressions of disgust and dislike, I felt as if I were seeing a stranger on the street looking at where a dog had just thrown up. They might have looked disgusted, but it didn't mean anything to me.

I also didn't feel the undercurrent of disgust in as many of the expressions. Somehow, the test had troubled me before TMS but it didn't bother me at all afterward. The TMS seemed to change how I perceived the expressions.

Then they handed me another test—recognizing expressions from photocopies of eyes, some of which were deliberately grainy. On that test, I felt a definite improvement after TMS. Shirley said, "Look closely. Take your time," yet I felt whatever reaction I had to the eyes almost instantly. Concentrating didn't change my responses.

I'd also done this test before that day's TMS session, with a very different result. Before TMS, I saw the pictures as just pictures, and I didn't get any real feeling from the eyes. After the TMS, my reactions were strong and clear. When I looked at each pair of eyes I knew right away what they were saying. I can't say how much better I scored, but it felt quite a lot easier.

With the tests finished, Lindsay and Shirley started to talk. As our conversation unfolded I realized it was easier and more natural to look both of them in the eyes. The feeling that I was spying or intruding was gone, and both Shirley and Lindsay agreed that I gazed at them more directly. I had usually instinctively shied away from direct eye contact, and I didn't anymore.

That was a little surprising, because the original stimulation hadn't hit me that way immediately. It had taken some time to build. "We'll see how I'm feeling tomorrow," I said to Lindsay on my way out. "Right now, I am not feeling like I am seeing deep inside, but that took twelve hours to develop last time." I went down to the garage, got in my car, and headed for home.

By the time I reached Worcester—the halfway point on my journey—I was starting to feel a little drunk. Just west of Sturbridge I found myself behind a state police car, and I dropped back, fearing I might get busted for driving while intoxicated. The internal dialogue that had quieted in the lab was working once again, as the little voices in my head asked, *Are we weaving back and forth? Are we going the speed limit?*

Needless to say, I made it home. Even though the world was moving like a ship at sea, I retained my ability to navigate. And as strange as my head felt, my reflexes and coordination didn't seem to be impaired. But I don't really know if that's true, because drunk drivers make the same claim, and they still crash. Later, when I told Lindsay about this feeling, she asked if I thought they should not let people drive home alone after TMS. I truly could not say whether my post-TMS driving ability was impaired or not. As she pointed out, the fact that I hadn't taken any pills or drunk any liquor didn't mean I wasn't in an altered state of mind.

There was no telling what would happen next, but the faux drunkenness I was feeling made me remember the hallucinations I'd had in April. Was something similar in store for me that night? It almost made me afraid to close my eyes. But bedtime was coming, and I'd have to do just that. When I did, the world started to spin the way it had before, but the hallucinations didn't come. Instead, I lay in darkness with a feeling of motion, as if I were sleeping on a sailboat at sea, rocking and moving with the swell. I wasn't getting seasick, but I wasn't falling asleep either. After a few minutes, I got out of bed, went up to my study, and composed an email to the scientists while the thoughts were still percolating in my brain. I reported the dizziness and said, "It may be a while before I fall asleep, but at least I'm prepared." I looked at my words later, and asked myself, *Prepared for what?*

I wished I had someone to talk to. Martha was there in bed, but sound asleep. The TMS had driven a wedge between us, and she wasn't eager to talk about what she called my "artificially induced feelings," especially late at night. I wondered how long we could keep things together. The recognition that I was failing at marriage once again weighed on me like a rock in the pit of my stomach. I'm sure Martha shared my fears, but we were both afraid to talk about them and they just grew worse in silence.

Returning to the bedroom, I decided to listen to some music through my headphones. When I did, I realized that the auditory clarity I'd experienced in April was back—but not the hallucinations—and with it, something new. The last time the music had come to life I had cried at the intensity of the emotions elicited by the words and melodies. The feelings that I'd experienced had been overwhelming. Now, that same thing was happening, but the music wasn't making me cry. Tonight's feelings were calmer, less dramatic, and they covered a wider range. There was joy and sadness in equal measure. And there was a new sense of connectedness. That made me wish I could reach out and speak to the musicians. I could not recall a single time earlier in my life when hearing the songs made me want to reach out and talk to the singers—even when they were right there beside me backstage.

A few minutes before, I'd turned on the music to put myself to sleep. Love songs and gentle jazz typically had that effect on me, especially jazz from the fifties and sixties—Stan Getz, John Coltrane, or Miles Davis. Now I was wide awake, fascinated by what I was hearing and feeling. What would I have said to those musicians had time travel been possible? Back when I worked with performers, my comments were mostly limited to how we adjusted the equipment, how the system sounded, or which guitar we'd use on which songs next set. I never offered comments on the beauty of their playing or how it made me feel. Now I wanted to talk to someone about the message of the song, not how the amplifiers delivered it.

I got back out of bed and trudged up to my study again, determined to capture a written record of my experiences of the evening. I brought my music with me, and as it played I looked at old photos that were stored on my computer. As familiar as they were, they also seemed changed. It took a moment to put my finger on the difference.

When I wrote Alvaro to describe the experience, I said:

> You're probably wondering what this feels like. . . . It's as if the audio processor in my mind just got upgraded. I listen to old familiar music, and it's similar, but richer. I notice things. At the beginning of a song, four bars in, the drummer misses a beat. Later on, I perk up my ears as I think, is that a xylophone playing back there? And indeed it is. Later on, I notice a little four-note riff on the bass. The individual components of the performance are somehow easier to pick out.
>
> When I speak, it sounds as if I'm listening through a sound system where the upper midrange bands on the equalizer have been turned up. My voice is crisper and better defined to me. To others, as best I can tell, I sound the same as always. Sight and sound is just, well, richer.

I went on to describe how color had become richer for me too.

> It's everywhere I look. Every color I see is now multi-hued and textured. Even the basic blue background on this computer monitor is different. The shading in the corners of the screen is obvious tonight, yet I never noticed it before. I can walk back to my bedroom, and in the gentle night light I see a hundred shades of color on the peach walls. I'm sure those subtle shadings were there all along, but I never saw them till now.
>
> One can argue that I could have seen any of these things if they were pointed out. That's not true. Look at the change in how I hear my own voice. It's not just a matter of paying atten-

tion. I'm hearing more definition. It's as if everything is sharper and crisper. You have to actually hear such a thing to truly understand it. As an analogy, it's a bit like wearing earplugs for quite a while and then taking them out. That is what I'm experiencing now, but in a more subtle way.

The sense of wonder at my musical insight kept me awake for many more hours. Dawn was breaking when I finally fell asleep, and I was late getting to work the next day. When I arrived, there was no eureka moment as I looked in someone's eyes, but I felt different just the same.

I didn't feel like I was looking into the souls of others, as I had back in April. I just had a greater sense of comfort and feeling of connectedness, especially when customers started to describe problems with their cars. That morning I found myself sympathizing with our clients and asking how they felt—with no prompting. The feeling this time, five months later, was more settled and less dramatic than it had been in April. The emotional insight this stimulation had built up was more like an old friend than a stunning new revelation. I'd told the scientists that this result was more subtle, but I gradually came to see that "more subtle" might be the key to what lasts in life.

I thought of Carly Simon's song "The Stuff That Dreams Are Made Of" and wondered if TMS was revealing the world I'd always dreamed of, right before my eyes. Then I felt sad, because her song was about the realization that the dreary partner beside her was really the special one she'd been waiting for, and of course the opposite was happening to me with the state of my marriage.

With no better answer, I put that sad thought aside and listened to the old music with ears that felt renewed despite the ravages of middle-aged deafness. A part of me felt twenty-one again, and that made me smile.

Different Kinds of Success

I MONITORED MY behavior closely in the days and weeks after the August session. And I had a lot of help—my family, my friends, and the scientists in the lab were all watching me and offering opinions. The thing is . . . I'd gotten what I'd long dreamed of. For a brief moment, I'd felt sure I was seeing into the souls of other people, and the different way they responded to me bore that out.

What I never for a moment considered is that I might be devastated by what I would find. My childhood had been pretty rough, with a violent, abusive dad and a mentally ill mom. There were a lot of hard parts, but I've always believed many of them slipped by me unnoticed. More than one psychologist has suggested that autism shielded me from the worst of my upbringing by making me oblivious to what was happening around me. With greater emotional insight, I lost that protective shield, and at moments it was devastating. I had created a fantasy that seeing into people would be sweetness and love. Maybe there was some of that, but there was a lot more fear, and jealousy, and anger, and every bad thing I could imagine. Sure, there were some good emotions, but they were very much in the minority.

Before TMS enlightened me, I thought the reason I often felt some-what down was that I could not receive positive emotions from other people. Now I knew the truth: most of the emotions floating around in space are not positive. When you look into a crowd with real emotional insight you'll see lust, greed, rage, anxiety, and what for lack of a better word I call "tension"—with only the occasional flash of love or happiness.

It's funny how that worked out. Before doing TMS, I was often anxious, and I reasoned that it was because my autism made it hard for me to un-ravel another person's feelings or intent. But now that I could read others more clearly, my plight was not necessarily improved, because one of the strongest emotions I sensed in others was their own anxiety.

I had recently started to notice what I called a "weird vibe" when I was around some people. First it was a sense I got from a few friends, like Richard, but then I realized I was getting the feeling from some customers too. Sometimes it began with a seemingly innocuous encounter. There is one that sticks in my mind even now.

"Would you like a ride to work, and we'll bring the car back when it's done?" I was asking a local doctor if he wanted his Mercedes sent back to the hospital where he practiced when his repairs were done. Most times people appreciated pickup or delivery of their cars, or a ride back to work, but this time I got turned down.

"No, my assistant is coming. I'm sure your time is *much* too valuable for that." A few months ago that comment would have gone unnoticed. It was not overtly nasty or critical. But with my new awareness I recognized it for what it was—a subtle dig at our company or me. I felt sudden deep sadness, as if the fellow had just announced he didn't much like me. The feeling was the same one I'd gotten as a kid, when no one wanted me on his team.

With a start, I realized that I had just experienced the adult version of rejection on the playground. I wasn't sure why it had happened or what to do. Why would he leave the car here for service if he didn't like us? The answer remained a mystery, so I did nothing. The man went to work, and we did the service that was requested. And all day long I reflected on those few words of rejection. It wasn't just the words—it was the way he'd said them, his attitude . . . a bunch of signals I never would have noticed before but couldn't forget now. How I wished he had never come in! It took all my self-control to keep quiet and not tell our service manager to call him and say we could not service his car and we'd leave it in his spot at the hospital.

I remembered my grandfather's words: *Sometimes it's better to be dumb.* Then I had a thought: *I'm the owner here. We have a thousand customers, and most of them are great. But a few are pretty nasty. They come in here bad tempered and complaining about everything we do. For all I know, they spend their whole lives bitching and being miserable. I can't control how they feel or what they do elsewhere, but I don't have to let them be miserable in my shop.*

In a spiral of negative feeling I let my mind wander to other unhappy customers and the things they had said. I suspected some of them were just unhappy people, no matter what we did for them. For some people, everything was too expensive, took too long, or there was something else broken.

For a moment I asked myself if they were right. We certainly took too long on some jobs, and try as we might, we inevitably made errors. All of us had a role in creating customer dissatisfaction at times. But most of our clients were happy with our work. Our surveys said the vast majority of jobs were done correctly, with a good level of quality. Sure, in the past I'd known that some people were never happy, but it didn't mean anything to me. They were never my favorite people, but I tolerated them just as I tolerated most everyone else.

Standing up to bullies was a good strategy when I was a kid, and I resolved to stand up for myself now. Things played out very differently the next time that doctor showed up expecting service. I listened to his requests, which were followed by a comment about making sure to "watch my technicians" that I found particularly offensive. We had never, as far as I knew, made any errors in servicing his vehicle, but he complained and warned me every time.

"You're a smart guy," he said with a smarmy smile. "But you're not the one fixing my car." This was followed by a five-minute rant about the shortcomings of my technicians and how they didn't quite know how to diagnose a car like his.

I didn't get upset. I just answered him calmly. "You are never happy with what we do. There's a Mercedes dealer in Hartford. It's less than an hour from here. Why don't you go down there and see if they can't meet your needs better? I'm not going to do any more work on this car."

Whatever outcome the doctor was expecting, that wasn't it. He back-pedaled quickly; now it seemed we were not all that bad. But my position didn't change. "We have a limited amount of time here, and I want to spend it working for people who appreciate what we do. If you're not happy with us—and that's obvious to me—go somewhere else."

Over the next six months I sent quite a few miserable people down the road in that fashion. "Why did you throw him out?" the guys in the shop would ask me. "His money was as good as anyone's." It wasn't about money for me. It was about how they made me feel. "We're not here to let customers treat us like doormats," I said.

Car repair is complicated, though many motorists assume it's simple. Things aren't always easy, and they don't always go as planned. Jobs cost more than people hope. Some customers don't want to believe what the carmaker or I say about maintenance. I point to a page in the owner's manual and say, "That's the oil rating your car needs," and they say, "That's not really true." I ask why they doubt the carmaker's specifications, and they get huffy and say, "Don't you understand that the customer is always right?"

That was a dangerous question for a customer to ask me, because it spoke to the heart of who I was. There was plenty I could not do, but I was absolutely confident of my gift for reading machines. Someone who came in asking me to fix his car clearly did not share that gift. If he did, he wouldn't need me to tell him what was wrong! If the customer was always right there would be no need for professional diagnostics or a car repair shop in the first place.

Liberace used to say he played classical music "with the boring parts left out." We've sort of done the same, but with cars. We now focus on difficult diagnostic work no one else can do and restoration that is as much automotive art as mechanics. As I watched this process unfold I thought back to my vision at the company's founding. I'd said we would work on cars people care about, and we do. But though it might sound funny, in all my past encounters, I had considered the needs of the cars first. Now, though some customers annoyed me, I was connecting better with most of them and striking a chord I had never before hit. Instead of being 100 percent logical when a client would come in with a problem, I began speculating aloud about how that person might have been feeling in her moment of vehicular crisis. "That must have been scary," I would say, and like as not, I'd get a response explaining exactly how scary it was when the car quit running in the left lane on the interstate in the midst of rush hour traffic.

"I'm glad I found you guys," a client would tell me, even before we had fixed his problem. I heard praise like that more and more, and the only explanation I could come up with was that my behavior was making people feel more comfortable. Nothing else had changed at work. The staff and the shop were exactly the same. That was a remarkable realization—the thought that TMS had opened my eyes to outside emotions and that awareness was translated almost immediately into success at work. I slowly learned that I couldn't simply pick a brand—like Land Rover—and assume that all Land Rover drivers cared about their cars. They don't. It takes an ability to connect with the drivers, because one guy can be proudly passionate about his rusty Subaru while another driver can be essentially indifferent to a new Ferrari.

As my ability to connect with people improved I came to see that many of our angry and dissatisfied former customers had not been car enthusiasts at all. They were just people who wanted transportation to work, and it happened to be a Benz or BMW that did the job. I'd been trying to share an excitement and interest that they simply didn't feel. Now that I can read our clients better I am far more successful at building relationships with the real enthusiasts who are the heart of our customer base, and that connection has made life better for all of us.

So while my new awareness didn't make my relationships easier, it did make the majority of them more successful. I like to think that the experience in our shop is more satisfying than ever today, for all of us, and I'm grateful for the TMS insights that pointed me onto that path.

Rewriting History

ONE OF THE HARDEST things about my emotional awakening was the way it reshaped so many of my memories. It may sound crazy, but all too often, it turned formerly good memories bad. And there's no balance. There's not a single bad memory that's now turned to good.

Sometimes as we mellow with age, we think back on hard times and say to ourselves, *Yes, it seems funny today but it sure wasn't funny when it happened.* But that's not what I mean. I'm talking about scenes of daily life that had been written into my memory banks and recalled without discomfort for years that now took on a new, disturbing meaning. There were the memories of former friends like Richard that turned bad as I interpreted the remembered scenes in a new light. But there were other incidents too.

When I was a teenager, my mother began seeing a new psychiatrist and started spending days on end at his house, which was populated by an ever-changing cast of characters, some of whom were pretty strange. Most of the doctor's hangers-on were creepy, but there was one fellow I felt a real bond with. His name was Neil, and he was a patient of the doctor's who was also renting a room in his house.

"You should get to know each other," the doctor exclaimed in his booming voice, and I decided to follow his advice. Neil and I started hanging out and talking about cars and music. Neil was twenty-one—old enough to get into bars—and I was thirteen. He even had a car. We began driving to rock and roll and blues shows, and he would help me slip past the doormen, who were always on the lookout for underage drinkers. Once inside, he bought me drinks while I stayed in the shadows and took in the scenery.

As we saw the same local bands night after night, I began thinking about how I could improve their sounds. I wasn't an audio engineer yet, but the notion that I could become one was starting to grow in me. The first few times I was timid as I walked up to the sound men and made suggestions. They were real adults, while I was just a kid pretending to know something. Some of them dismissed me, but others tried my suggestions, a few of which even worked. That success gave me credibility, and it was how I got my start working with musicians. I moved on to fixing and then designing sound equipment, and in the space of ten years, I was working with some of the biggest bands on the planet.

But that was all far in the future when Neil and I sipped drinks at the Rusty Nail and other local dives. Back then I was just a teenager exploring a big new world. Everything was exciting, but there was a dark side, one I'd totally misunderstood until the TMS-induced changes shone a new light on my memories.

Neil and I talked a lot in those days, and one of the things we talked about was sex. That wasn't something I had experienced personally, but I'd heard stories and I wanted to know more. I fantasized about having a girlfriend one day, and what it might be like. I never considered the idea of a boyfriend, until Neil raised the possibility. He was a fairly educated fellow—at least I thought he was—and he liked to tell me stories about history. Many of those stories involved sex, and most of the sex was between men, though there was an occasional female in the orgy.

"What do you think sailors did, crammed together in the fronts of sailing ships?" he'd ask me. I assumed they ate, slept, and worked the rigging. It was a surprise to hear that the thing they looked forward to every night was the sex. And then there were the Roman centurions and the Spartans. Who could forget those great soldiers of antiquity, famed for their manly sexual feats? Neil was happy to enlighten me about the role of gay male sex among the ancient Greeks and Romans.

Eventually the suggestions became personal. He told me I was a handsome young man. According to him, lots of older guys would be honored to take me under their wings and teach me what I needed to know. I didn't completely understand what he meant, because guitar amplifiers were what I needed to know right then, and they weren't even part of his conversation.

Finally he offered to teach me himself. "We could go back to my place," he would tell me. "You've never had real sex, and you know you want to try."

"I don't think so," I told him. "I want to wait and get a girlfriend. Maybe I'm not the same as those Romans."

Nothing sexual ever happened between us. Neil never introduced me to the joys of orgies or Spartan conquests and I forgot those conversations as I found success in musical electronics and then with that girlfriend I had hoped for for so long.

I didn't think of Neil again for a long time, and I can't remember exactly what triggered my recall of that story, but this time—after TMS had awakened that emotional response system in me—I saw the whole thing differently. Now Neil seemed like a crafty pedophile, reeling me in by taking me fun places, all the while waiting for his moment to take advantage of me. Back then I thought we'd drifted apart when I got older and became more independent, but the truth is, he probably found another teenage boy to prey on. What once seemed kind and caring now felt corrupted and awful. It was like a negative version of the new, expanded way I heard music or the heightened expression of colors I now saw.

It made me sad to see the truth. But with a bit of reflection, it also made me question what the truth was. Was it the fun I had when Neil and I were together, or was it the imagined ugliness of his intent? And am I right to even say his intent was ugly? Maybe he was like me, just a bigger lost kid, looking for love. After a time I came to see that it might be all that; the fun

I had was real, and though it's easy to say he was an evil sexual predator, the truth is, I have no idea.

Thanks to TMS, my head is now full of turncoat memories, and I have to confess that I'm not the first one to realize what happened. A few months previously Kim Davies had told me that TMS had shown her everything that went wrong in her life and why her relationships had run into trouble time and time again. When I heard her, I assumed her perception was like mine, and we had a "glass half full/glass half empty" sort of situation. I told her that TMS had shown me the same things, and I now saw what I could do differently going forward. I believed that at the moment, and I still believe it today. But I now understand that that's not what Kim meant. What she meant is really what she said, and I know that because I feel it too—it just took me a little longer to understand what was inside me.

Everywhere I look, there are memories of things that went wrong. And thanks to TMS, I now recognize that many of the events went wrong because I failed to understand someone else's feelings. Kim saw the same thing. TMS may have given us a temporary ability to see feelings in others, but it's given us a lasting clarity in interpreting the memories of life experience.

I never asked Bob what happened to his daughter. . . .

I hurt his feelings when I failed to thank him, and that's why he never called back. . . .

I should have asked her how her day went instead of walking in and boasting out about the car I just sold. . . .

Individually, these memories are just little things. None of them are life changing. But the weight of knowing what I did wrong, a thousand times over, is heavy. When I remember things I said or did, I cringe and wish I could go back in time and undo my blunders. Still, we can learn from our failures and with any luck not repeat those mistakes as egregiously tomorrow. It's not easy. Knowing what I did wrong is not enough to undo a lifetime of learned behavior, and my tendency to behave the same way is still strong. Yet I'm doing my best to change.

It often takes me quite a long time to arrive at the more correct conclusions. Psychologists say autistic people have processing delays, and maybe this is one of them. Or maybe it's just a very complex problem. It's as if my emotional brain needs time to reflect before arriving at an understanding of what a particular scene or experience meant or means.

Nowadays this happens to me all the time, as things I see trigger memories of the recent past. The trigger itself is usually something innocuous; maybe I'll see a car that looks like one we had at work a few weeks back and the sight makes me remember what someone said or did, and the recollected scene plays itself back in my mind. When that happens, the meaning that had been confusing in the moment becomes clear, and I feel like I've been kicked in the stomach. *You blew it again,* the little voice in my head tells me.

It took me quite a few years to come to terms with my autism, once I learned about it. The disability side was obvious to me from the beginning; it was the gifts that were so much harder to see. TMS worked a little differently. I saw the beauty in music and color right away, and those things came to me like gifts dropped out of the sky. But there were also costs that took even more time to discover.

Fear

ALVARO HAD WARNED all along that the effects of TMS might not always be positive. I was changing—mostly for the better—and I'd been on a steady upward glide. For the first time in memory, I was free from the anxiety that used to torture me every fall. My first book had been a success in hardcover, and it was looking pretty good in paperback for the fall of 2008. But that all changed in the space of a few months, as the economy imploded.

There had been some problems on the horizon to be sure. My marriage was unstable, and I struggled with Martha's all-too-visible depression and my newly heightened awareness of it. My son was headed for trial over the chemistry experiments, and he might end up in prison. But the straw that broke the camel's back was the economy.

I'd lived through several recessions, and this one didn't worry me much in the beginning. I didn't feel a connection to the stock market or what pundits were describing as the real estate bubble. *What does it matter to us?* I reassured my staff. *We are car mechanics, not stock traders.* Being autistic, I had never felt connected to most of the goings-on in the outside world, except for those that directly affected me.

My business had taken off in the 1990 recession, as new car owners looked for less costly alternatives to the dealer. I initially assumed this recession would be the same. It wasn't. Old customers stopped coming in, and new ones didn't arrive to replace them. There were days I thought the office phone was broken. When I called former customers to see what was wrong, I heard one scary story after another. Overnight, people who had looked after their cars faithfully stopped caring for them at all. "We couldn't afford it and had to downsize" was a common refrain that winter. The stories in the newspapers got worse every day. Business sagged and then collapsed.

By the spring of 2009 everything had changed. The stock market had tanked and millions of people had watched their retirement investments evaporate, sometimes overnight. My own stock portfolio—never anything to boast about—had lost half its value. Customers who had been very attentive to their cars neglected them and then became angry when that resulted in breakdowns. Some were actually abandoning vehicles when they couldn't handle the repairs. Not only were they dumping their formerly treasured but now run-down cars on me, they were ignoring my phone calls and mail. When they did show up to service their cars, they were edgy and anxious, something I had not seen before. While it's true that I was oblivious to my clients' moods in years past, I still heard what they said to me, which was usually chatter about weather, kids, and other light stuff. Now they told me of their hard times: jobs lost and kids they could no longer afford to send to college.

The realization that I couldn't do anything to help my customers was very disturbing, and that was a new feeling for me. In the past, my response would have been to tell people that their breakdowns were a predictable result of inattention, and they should change their ways or they'd end up walking. Now I found myself feeling compassion as I realized some were in the position of choosing to pay for a car repair or last month's mortgage. I'd always heard that compassion was a good thing and that autistic people like me were disabled by its lack. *If that's true,* I told myself, *I'm better off disabled, because feeling the pain of others is dragging me down.*

Still, the troubles of our service customers were probably a good distraction for me. As spring approached, Cubby was headed for trial in Hampshire County Superior Court in Northampton. Of all the thousands of criminal complaints in our county every year, only about fifteen of them are serious enough to turn into full-blown superior court trials. Cubby's case was one of them. The other trials were for rape, murder, armed robbery, and arson. We still could not believe that the prosecutor had lumped Cubby in with people like that, especially since there was never a victim or a complaint. We were watching the court closely, and it was frightening to see one trial after another end in conviction. The Superior Court was looking more and more like the anteroom to the Concord state prison. Most criminal cases are resolved by pretrial negotiation. Not this one. My son was still facing up to sixty years and we were all on the edge of panic, wondering what was going to happen.

My search for good news—or something to buoy my spirits—was becoming desperate. The only feelings in the air seemed to be anxiety and fear, and all the news was bad. Before TMS, war might break out in Europe and thousands could die of disease in Canada, but I would be unfazed as long as none of tomorrow's service appointments canceled. Now everything I heard, saw, or read was fraught with feeling, almost all of it negative. Everywhere I looked people were scared. It was impossible for me to stay on an even keel in the face of such overwhelming unpleasantness, and I wished I had my autistic emotional oblivion back. Great as the TMS-derived insight was, it came at the cost of losing a powerful protection. Without it, I was essentially naked in a hostile world.

I expressed that sentiment to Alvaro, and he said, "That's a very astute observation. There is a body of research that shows we can 'catch' emotions from other people—principally anxiety and depression. That didn't happen to you because your autism acted like a protective inoculation. I feel sorry for your present distress, but at the same time I am thrilled that the therapy unlocked all these feelings in you. You're feeling a new kind of empathy, and it's surely a big strain to adjust."

Unfortunately, Alvaro's explanation and enthusiasm didn't make me feel much better. It gave me a momentary smile, but that didn't last. Life was just too painful.

My marriage was all but dead, as I kept failing to deal with my newly emergent response to the cloak of Martha's depression. I had been oblivious to it for so long, and I felt horrible because I'd turned on her without warning. There was nothing to say, because this wasn't something talk or even counseling could change. She thought our troubles were an incompatibility precipitated by science. Unpleasant as that was, it was magnified by the guilt I felt for letting her down; they say it takes two people to make or break a relationship, but in this case both of us were keenly aware that it was me and TMS on one side and her on the other. I was different and she was the same, and things had ceased to work between us.

Now that I was more able to read the feelings of people around me, I recognized my impatience with Martha's sadness and saw that my pushing her to do things she didn't necessarily want to do was mean. Before TMS I told her that having Cubby and me around was good for her because we wouldn't put up with depression. We'd insist she get out of bed and join us out in the wider world. After TMS I found myself looking in the mirror—metaphorically—and thinking, *What am I doing?* My expanded sense told me she wanted to be left alone, which must mean I was the one who wanted her out and about. If that was true—and my logical brain insisted it was—then the whole effort to get her out from under depression was just self-serving, and I felt ashamed.

Another thing that bothered me was realizing that I was driven to *do* something all the time, while Martha was content just to *be*, sitting quietly with a book or a glass of wine. No one had ever taught me how to relax, and the excitement of the TMS made me even more energized. But she wasn't like that, and I saw how I'd been jabbing at her constantly, trying to make her more like me, even as I wished I could sit quietly and relax like her. My behavior was making her depression worse.

I thought about it all as I listened to Diana Ross sing "Touch Me in the Morning." Thirty years before, all I would have heard in the music was a river of golden sound pouring through the crossovers, limiters, and amplifiers. All music was beautiful when the sound system worked and horrible when it didn't. Now, I didn't even notice the sound gear. She sang the lyrics and their meaning hit me like a punch in the stomach, as if I'd never heard them before.

We don't have tomorrow,
But we had yesterday.*

My mood took a brief upturn when Cubby was found not guilty on every one of the prosecutor's overblown charges. We could call them ridiculous now, having won, but they'd been very serious until that moment. But even that wasn't enough to lift my mood for long. And Cubby didn't stick around to celebrate. He hated the stress in the house and wanted to be on his own. At age nineteen, he was an adult. He moved thirty miles away to Vermont, enrolled in community college, and took up with a new girlfriend. He was off to a good start as a young adult—or at least a better one than he'd have gotten in jail—but I missed my little boy.

The economy remained terrible throughout the summer. Our revenues at the garage declined by 25 percent—something I'd never seen in twenty years in business. After almost a year of bad news, I began to lose hope. Until then I had been a generally optimistic person, assuming that tomorrow would be better than today, and events had generally shown that to be true. I'd had down moods and periods of depression, but I kept going because I still believed in tomorrow, even if I was sad or lonely. Now that faith was shattered. In its place was the realization that tomorrow would be worse, and next year would be awful, if I even made it that far.

That made me despondent, but it wasn't really depression. It was more like a combination of absorbed emotion and a reasoned response to what was unfolding around me. A few years of growth had made me successful,

---

* "Touch Me in the Morning" was written by Michael Masser and the late Ron Miller, and first recorded by Diana Ross in 1973. It was one of the songs I loved to hear performed live.

and I'd been a fool to think it would last. Now, with sales falling at a pre-cipitous rate, the only question was how long I should stay in the game. Was it time to concede defeat?

A year earlier, I'd thought of myself as a success. Now I was a failure. Whenever I tried to tell myself that was just unhealthy self-pity, my inner voice said, *No it's not! It's reality.* More and more, I came to believe that was true. And the endpoint of a worsening situation—I realized—was my own death. I began to think about that more and more as everything fell apart around me. I wondered what would happen. Would I lose every-thing and sicken and die? Or would I kill myself in a moment of misery? I tried to think positive thoughts, but I could not sustain them. Yet people still looked at me as a success, and at work they expected me to be a leader. That was the hardest part—going through those motions while feeling I was a fraud.

There were several times that summer when that voice in my head said, *This isn't going to get better. You should just end it now.* Looking back on that time, I think suicide was an impulse that snuck up on me when I was overcome by what felt like never-ending psychic pain. In one of those moments I came an inch from shooting myself on my back deck, but at the last second I turned away.

As horrible as my life had become, I must have maintained some shred of hope that it could get better. Things probably didn't look that bad to my friends and family, but they weren't in my head. Maybe the market would improve. Maybe business would pick up. My friends tried to reassure me, saying business wasn't that bad (it was), everyone else was in the same boat (some were and some weren't), and that I was more than just a living representation of the money I had in the bank. But I disagreed. I had known what it was like to live on the street at sixteen, and I told them I'd rather kill myself than be on the street again.

Things didn't get better that fall, but they didn't get much worse either.

As winter closed in I realized I had to make some changes. I knew that I could not stay married to Martha; her depression was overwhelming me. But I felt horrible for even having that thought. She was depressed when we'd gotten together, after all, and I tolerated it fine till TMS changed everything.

Finally, I told her I would be moving out. That Christmas Eve I drove my car to the boat launch ramp on the Connecticut River. I sat in the car for an hour, looking out at the dark water flowing past in the light of my headlamps, wondering if I should drive into the river and end it. I felt I had failed everyone important in my life. I'd been oblivious to my son's autistic detachment and allowed him to carry out dangerous experiments that had almost sent him to prison. I'd blindly pursued the TMS in the self-centered hope of making myself better, but the real result was that I felt the full force of Martha's depression and I couldn't handle being married to her anymore. And with the economy in a shambles my company was on the brink of insolvency. If it failed I had nothing left. There didn't seem to be much to live for. The logical next step would be to floor the gas pedal to get the car far enough into the river so that it would sink. I had good insurance. It would take care of the people I left. At this point, the money would do them more good than I would alive. Yet I hesitated, just as I'd hesitated earlier with the gun.

I sat there in the dark, hurting. I wanted so much to make the hurt end. Being dead would end the pain, but it would end everything else too. And I wasn't quite ready to do that, because some part of me knew that things might not be truly as bad as I imagined. Even as one voice in my head told me I was doomed, a quieter sense of reason remained to point out that it wasn't over yet. There was still money in the bank, and I still had a reputation. Recovery might not be looking likely, but it was possible.

Time passed silently as I looked at the water, flowing by in little swirls and ripples. Putting the car in reverse, I backed up the launch ramp, turned around, and drove into town. I went to a local restaurant, where the economy had ensured there were plenty of empty seats, and ate a solitary dinner. Then I drove to Springfield and checked into the Sheraton hotel. I had decided to start again, though what I was starting wasn't completely clear.

Movies and books have portrayed life in a hotel suite as a grand and glamorous thing. If only that were true. My room may have been comfortable, but it was unremittingly lonely. The restaurant didn't offer much solace. There were some enjoyable dinners with the owner and his friends, but when the night was done they had families to go home to, and I was left to trudge upstairs alone.

Martha and I went to mediation and agreed to divorce while continuing to own the car business together. I hoped that would work, and so far, it has.

I began to imagine myself as different, and I fantasized that the social people who'd spurned me when I was younger might embrace me today. *I might even marry someone like that,* I thought. That notion led to dating disaster as I learned—quite painfully—that my enhanced sensitivity did not allow me to predict who would be a true friend, supportive and trustworthy, and who wouldn't. After one woman left me I looked at some pictures of our time together. When I did, something peculiar emerged. In photographs her facial expressions gave off a clear message: she did not want to be there. In hindsight, I saw that while her words had been delivering one message, her face revealed another, and that reflected some inner conflict in her mind. *Better that I got out when I did,* I thought. TMS was instrumental in helping me come to that realization, even though it had not allowed me to see the mismatch in the first place. Before, I would not have sensed anything from the photos either.

The promise of a new relationship had lifted my mood, but its failure a short while later knocked me right back down again. That left me in a pretty awful state, and people at work began commenting on my behavior. "Are you having a problem with drugs?" our service manager asked me one day. She had watched my moody ups and downs—which I had never exhibited to that degree before—and concluded they must be the result of drug addiction. I assured her that I was not, but it took her some time to conclude that I was telling the truth. Her question rattled me and made me wonder how messed up I looked to the outside world.

"You get angry with people now, and you never used to before." That was what Martha told me when I asked her opinion. I realized she was right. I used to bottle up all those feelings inside me; I didn't know how to get them out. The result was that people assumed I was unflappable and on a permanent even keel. But all that had changed.

A New Beginning

THE FOLLOWING FALL, with my divorce behind me and dating nightmares in the background, I ran into Maripat again when she brought her car to the shop. Quite a lot had happened since the last time we spoke, and I filled her in on all of it as I gave her a ride to work. I told her about moving out of the house, filing for divorce, and taking up residence in the Sheraton downtown, and then, once the divorce was settled, ending up back in the house, though now I felt kind of lost in that big place by myself.

It was out of character for me to have opened up like that; I don't usually go on about my personal life with service customers who stand at our counter. But she was a friend too, not just a customer, and I guess I sensed there was something different about her that day.

To my surprise, Maripat told me she'd broken up with her boyfriend. She was quite surprised to hear I'd gotten divorced. And I was just as startled to hear she was unattached, since she'd always had her boyfriend follow her to drop the car off. But this time she didn't have anyone with her, and she asked for a ride downtown. And at some point during the ride I got up the nerve to say, "We're the same age, and relatively healthy. . . . We should go out on a date."

A few days later we went to dinner at a Chinese restaurant in the Berkshires and talked for most of the night. We had many things in common, much more than I'd realized. We'd both dropped out of school and we'd both had alcoholic parents. Both of us left home as teenagers, and we'd educated ourselves as we went.

We were both very driven at work, but outside the office we had some differences. She was into yoga, meditation, and other ethereal things. She knew little of machines and engineering, but as she pointed out, I knew little about spirituality. Both of us liked hiking and the outdoors.

Things looked and felt promising, but one test for people our age is what the kids think. Maripat had three children—a teenage son who lived with her, an older son who lived with his dad, and a daughter who was on her own in New York. We were both relieved to find the kids quite supportive—even Cubby. When we got together she actually helped mend my wounded relationship with my son, and with my mother. I thought that was really neat.

On top of that, I am an eater and she is an excellent cook, so food has always played a role in our relationship. Maripat is also a very family-oriented creature, which meant we involved the kids in our dining. She began a tradition of cooking Sunday dinners and inviting all the kids and their friends. They didn't all come every week, but at least one or two were usually there, and sometimes the whole pack.

We got married the following summer.

When our first Christmastime together arrived, she prepared for an even bigger crowd as friends from the neighborhood stopped by, and her children invited friends from town and school. Cubby brought his mother, and I was amazed by what happened next.

Maripat and my first wife became best friends. In my self-centered way I assumed I was the only thing those two would have in common, but I was wrong. From the beginning, they shared many interests. Sometimes I thought the two of them were more compatible with each other than either of them was with me!

Maripat invited Cubby's mom to dinner every week, and they jostled and joked, often at my expense. There were moments that made me uncomfortable, but it was the best thing possible for my son, having Mom, Dad, and Maripat all there together. My relationship with my first wife had always been tense, but that melted away when Maripat came on the scene. It was the most remarkable thing.

Through some stroke of great fortune, Maripat's kids actually embraced life with me. Her son Julian, who lived with us at the time, used to help me out around the house every day, and his brother, Joe, would pitch in when he visited. Lindsay, away in New York, always welcomes me, and I know what a gift that is, after seeing other mixed families where no one gets along.

It was amazing what she did to bring our new family together.

With Maripat in my life, and a gradually recovering economy, I returned to a level of psychic equilibrium. I was certainly scarred by the events of 2009, but things were once again looking up. The car business survived the downturn and was regaining the ground that had been lost. I got back to writing and dove even deeper into autism science.

I'd taken the first steps in that direction shortly after the first TMS study, when I agreed to review autism research proposals for the National Institutes of Health. Now I was part of several research review boards. I'd started working with the International Society for Autism Research (INSAR), the professional society for autism researchers, and they had named me to two committees. I was lecturing at colleges and universities on autism and speaking at conferences and schools about my life experience as an autistic person.

Much of that work had come my way since the TMS project began. Good as that seemed, it also felt overwhelming, and I started to panic, wondering how I'd do it all, and run the car business, and take care of my family.

Maripat made all that possible by turning my empty postdivorce house into a home. Her creation of a safe place I could return to makes it possible for me to venture out and do the things I am called on to do today.

Then something even more remarkable happened. After observing the lecturing I was doing, Maripat joined me on the road—not as a travel companion, but as a partner in reaching out to neurodiverse people and their families. We began doing joint workshops on life with autism, and she was immediately swamped with people who wanted her perspective.

Her first presentation received a standing ovation. She turned out to be a really good speaker, with a message the audiences were eager to hear. As I embraced the science, she took on the emotional issues. Her counterpoint to my logical reasoning made for a good combination. Both of us together became much more powerful than either alone. That was a new experience for me, one that is still unfolding.

Tuning Out the Static

IN THE SUMMER OF 2007 everyone in my family had been thinking hard about how to promote my first book, *Look Me in the Eye,* which would go on sale that September. My brother, Augusten, got the idea that we should make a video, and he walked down to the meadow where I was working with my tractor and accosted me with his camera for an interview. The resulting "tractor video" ended up on YouTube and other websites, and it attracted a lot of attention. At that time, there were very few videos out there of autistic people talking about anything, let alone a memoir.

It all seemed pretty ordinary to me, but I guess the responses I gave to my brother's questions were not what the average person expected. (You're welcome to go online now and watch it yourself.) Some of the comments that viewers left were funny, and others seemed hopeful. A few hurt my feelings. One commenter in particular had said, "He looks and sounds like a talking robot," and I burned with shame.

My book came out, nine months passed, and I forgot all about the video. Then the doctors did the TMS experiments on me, I began speaking widely, and new videos appeared, with new comments. Now the tone was different. "Wow," people said. "You sure have changed. Look at the comparison between that tractor video and the film of you now! What a difference!" When I went back and looked at the film my brother had made, I saw it in a whole new light. I realized I was totally devoid of expression or animation throughout the entire clip. Hurtful as it seemed at the time, "talking robot" really was an accurate descriptor. What's remarkable is that I'd totally missed that before, but I got it instantly when viewing it after TMS.

Interestingly, when I look at the video now, I don't feel hurt by the comments. Instead I marvel at how much I've changed and how fast it happened. When I look at myself in videos made after TMS there are smiles, eyebrows lifting, and many hand gestures. None of those things were visible in the earlier tractor video.

Halfway across the country, a man in Minnesota watched the videos and read my online musings about TMS. Timothy Taylor was the editor of an academic journal and the father of a teenager with autism. He'd come across my posts by chance while researching another topic. Fascinated by what he saw, he shared the videos with his wife, Kimberley Hollingsworth Taylor.

Nick, their son, was a lanky eighth grader, kind and smart, with a large vocabulary and A's in most of his classes.* With diagnoses of autism, attention deficit disorder (ADD), and obsessive-compulsive disorder (OCD), he was still doing a lot better than I ever had in school, but his good grades didn't insulate him socially. There, he ran into the same challenges that I had at his age. As Nick moved through junior high, he took to saying he was "just not a friend person," spending his free time playing Minecraft and watching YouTube videos rather than hanging out with other kids. His parents watched him struggle to connect with other people and to function in his daily activities.

As Kimberley described, "Although Nick is very smart in terms of vo-

---

* This chapter is based on my correspondence with Nick's mother, and most of the dialogue is quoted directly. I've changed Nick's name. His mom, Kimberley Hollingsworth Taylor, and the Clearly Present Foundation are identified by their proper names. Her foundation continues to support TMS research.

cabulary or math skills, it took him much longer than others to complete class work and homework and activities of daily living. He wrestled with OCD-type compulsions—he would write, erase, and rewrite letters over and over until they looked 'perfect,' so writing a few sentences or a few math equations could take an hour." As time passed and Nick got older, he seemed more and more "stuck." In conversations, or when asked about anything, even about activities or topics he'd enjoyed in the past, his favorite word became "no." Any change in his routine could become a high-stress event for the entire family, including his brother and sister. Nick's tendency to stutter and repeat the same phrase or question also seemed to be worsening as he got older.

When his parents watched my videos and the changes in me, they saw hope—that something similar could be possible for their son. Nick's mother tracked Lindsay down through Harvard, only to be told there were no current studies involving children. "I'll keep your name in case we start a teen study," Lindsay offered. Meanwhile, Nick continued to struggle with school and social life. He tried ADD medications, but those didn't really help with his core social issues.

When I was a teenager, the thing that had helped was simply growing older and having people around me recognize my value. Nick was on the same track I had followed, and that's a scary place to be in the beginning. When you don't have friends and can't do what people ask of you, it's very hard to imagine that things will magically get better in the future.

If medication didn't help, and counseling didn't do much, what else was there? It took a few years, but Lindsay eventually called Nick's parents back in the spring of 2012 to tell them the lab was beginning a study to look at the short-term effects of TMS on teens with autism at Boston Children's, another of the Harvard teaching hospitals. After talking with Lindsay and the other researchers, Nick decided to give it a shot, and over spring break, he and his mom traveled to Boston to meet the researchers and learn more.

In the years since I had done TMS they had refined the technique considerably. When I did the study, the scientists stimulated me with one pulse a second for half an hour. Nick would receive a much faster series of pulses, called a "theta burst," and the session would be over in a minute. Lindsay and others had found the bursts of TMS to be more effective in other studies. The target for Nick was Broca's area behind the right front temple, the same location that had produced such dramatic results in me four years earlier.

The problem with theta burst TMS is how it feels, as the rapid *bap-bap-bap* gives the distinct impression that something is drilling into your head. I'd experienced that in a 2009 study and described it as "kind of nasty." There was no sense of the meditative calm I'd experienced before; the rapid pulses just slammed into me one after another, ten times a second. "You'd have to really believe in the promise of this to sit through something like that repeatedly," I'd told Lindsay at the time.

Nick's stimulation was supposed to consist of six hundred pulses over forty seconds, but he asked them to stop three-quarters of the way through. Lindsay didn't know if they'd done enough to make a difference, but she hoped to find out with before and after exercises, much like the ones I had done. She asked Nick to look at pictures of eyes and choose the word that matched the emotion the eyes were expressing.

His numerical score showed a small but statistically significant improvement, his mom told me. "But the real change was in how he did the test. Before stimulation, he wrestled with what he was seeing, and it took him half an hour to answer the questions. After TMS, he breezed through the test in just eight minutes. Then he looked the researchers in the eye, thanked them, and said goodbye. On the way out, he walked beside me all the way to the elevator, which was a big departure for him. He didn't feel any different, but I saw a change in his behavior right away.

"He was actually walking differently when we got to the airport to go home," Kimberley marveled. For the first time his mom observed him matching her pace, and that of other travelers. He stayed beside her and kept out of other people's way. His mom believed he was sensing the people around him, for the first time in her memory. "None of the drugs or therapies we tried over the years ever accomplished that," she would tell me later.

When they got home his family saw even more changes. "We went to a big indoor water park, where his younger sister ran up to him, excited to show him some of the water slides. As they walked away and she talked and gestured, I noticed him walking right beside her. He also turned his torso, shoulders, and head toward her, paying attention to what she was saying. Subtle as that was, it was a totally new sign of connection to someone else."

His parents observed other new connections to the people around him, and they talked with Nick about them. Nick described the change himself as going from "looking at the world through a static-filled TV screen, and after TMS the static is gone and things are just clearer." If only it had lasted! Lindsay had warned them that the main effects would be temporary, which I'd seen for myself. Seeing those changes fade over the next few days and weeks was surely very hard for Nick and his family. They looked for someone who could do more TMS for Nick out in the Midwest, but there was nothing available. A year later they returned to Boston for a follow-up stimulation, but like my own follow-up in the summer of 2008, the results from the second session were a lot more subtle. Kimberley and her husband had asked Lindsay what more could be tried, and her answer was to invite them back to Boston in the spring of 2013 for ten weeks of twice-weekly TMS.

I first met Kimberley that spring, just a week before they were scheduled to fly to Boston. I was speaking at the public library in their hometown, and she came to the event. Kimberley walked up to me after I spoke and described their family's journey, which paralleled mine in so many ways. I said my own TMS experience had been one of the most important things to happen in my life, and I wished them luck in their travels. I resolved to stay in touch to find out what happened.

After the ten weeks of biweekly stimulation in Boston, Nick's parents still hadn't seen any sudden or dramatic shifts in their son, the way they had after that first session. Nick said he felt about the same, but over the summer that followed, all of them saw real but subtle changes. For example, Kimberley told me that Nick took the next year's English requirement as a summer online class to make the following year easier. "The course covered a year's worth of material in eight weeks, so the pace was roughly equal to doing a week's worth of English assignments each day. He did that successfully at home with minimal supervision, and he couldn't have done that before."

The pattern of Nick's second series of stimulations felt familiar to me. My own experience with restimulating an area had been much the same. The second time around, the effect was more subtle, but more and more changes became visible as time passed.

Kimberley reported that Nick's "ability to choose," which he'd shown after the first session, was back. Where before he got stuck on tests or puzzles, he now made choices easily and moved through the work quickly. He was more open to going places and doing new things. That made him better able to negotiate daily life. That summer, during a family trip to Montana, he said, "It's like there is more of me here now."

My own son had said something very similar when he described a sharper, clearer world around him, and I'd experienced the same thing myself. Kimberley said, "Nick used to have a one-track mind, but now he participates in conversations about things outside his usual interests. He also has more insight into his own reactions. For the first time he recognizes his feelings and discusses them."

When Nick returned to school that fall he seemed transformed. He was much more attuned to those around him, both teachers and students. His mom gave me an example: when Nick's teacher told him she had to add more material on economics to her classes because the curriculum had changed, and that she found this a bit challenging, Nick turned to his mom and said, "We should give her Dad's economics book. That might be helpful."

"That comment might not seem like a big deal," she told me. "But for Nick to understand something outside of himself, to make the mental connection between her need and a resource his dad had, and to suggest helping in a way that fit with the flow of the conversation—that was a remarkable thing. Nick had always been empathetic if his attention was focused on a certain person, or on the characters in a book or movie, but this was different. He was reacting to an ordinary person in a conversation outside his usual set of topics with a productive suggestion. And he made the comment only once, in a way that fit in with the rhythm of the conversation, with apparently no need to repeat it over and over."

Later that fall his teacher wrote a note to Kimberley about how well she thought Nick was doing. "He spends less time balancing on his tippy toes and seems to stutter and repeat himself less. He also carries himself better." And when he left to look for a new classroom, he was able to connect with another student who showed him the way.

As the school year progressed Nick's gains continued. He was getting his homework done by six, and he solved ten math problems in an hour. Prior to the TMS he'd averaged two. At home he described conversations he'd had at school and even remembered the other kids' names. Best of all, a group of four girls asked him to attend the homecoming dance.

Sometimes he found the changes overwhelming. "People I have never seen before and have no way of knowing me suddenly say hello." He suspected that his mother had somehow persuaded strangers in his large public high school to be nice to him, and this made him angry. His mom said, "Honey, if I had that kind of power I wouldn't have waited until this year to use it!" One day, he said in frustration, "I have ADD and it's really distracting to have people talking to me all the time and to have to see all these faces all the time."*

But then his gains started to fall off. He became gradually slower in getting homework done, needing more prompts and reminders. The OCD symptoms returned, further slowing his schoolwork and making it tough for him to throw things away, or fill in bubbles on tests, or even write things down without erasing and rewriting. At one point he told his mother, "Yeah, those girls who asked me to the dance have figured out that I'm different and now they don't want to have anything to do with me."

"Our family has been living our own version of the sad parts of *Awakenings,*" Kimberley told me recently, referring to the 1990 movie based on the book by Oliver Sacks, which tells the real-life story of finding a drug

---

* Studies have shown that some 40 percent of people diagnosed with autism also have ADD, ADHD, or OCD.

that awakened patients from catatonia—but the effects of the drug then faded.

As his family watched Nick regress, his "stuckness" returned in full force, and the normal wing-spreading recalcitrance of adolescence made things even worse. "Today, Nick says he can't remember anything being better after the TMS, he doesn't acknowledge ever making any positive gains, and he professes himself unwilling to try it again. The school assignments he'd started breezing through are once again an insurmountable challenge. His ability to participate in conversations with others has slipped away, and he's back to being interested in little besides Minecraft and YouTube videos."

As Kimberley says, "Outside of gaming and a few other interests, he doesn't participate much in conversations unless we drag him in. He doesn't ask us how we are, and at mealtimes, he sits with his face and body twisted away from us. The lovely parts of his personality that appeared after TMS are hidden once again."

One part of me hears that and thinks, *Nick's doing what he wants, and who are we to suggest that he change?* Yet I've been there myself, and I know the balance between wanting to pursue my own interests and wanting to be part of society. For most people, gaming and withdrawal are not paths to independent living. Teens like Nick have to make some hard decisions, decisions that might be easier for older people like me, with a greater foundation of life experience.

Nick's mom described something else that really echoed my experience: "We could have taken Nick to social skills classes for the rest of his life and he wouldn't have learned what was unlocked inside him with forty seconds of brain stimulation." It all made such an impression on her that she founded a nonprofit organization called Clearly Present to promote the development of TMS as a therapy for autism. Last year she held her first conference, just before the annual International Meeting for Autism Research (IMFAR). I hope she is successful at raising awareness among the folks who fund medical research because I fully agree that TMS has tremendous potential.

When we talked, she reminded me of Atul Gawande's thought-provoking *New Yorker* piece called "Slow Ideas." In it, Gawande says that we all want useful medical innovations to spread virally, and in a few cases they do. But for the most part, for large numbers of people to benefit from a new medical discovery, it takes twenty to forty years between the original inspiration and widespread deployment. Gawande's article spoke to both of us. If others could share our experiences, there would be a lot more interest in supporting this work.

Nick's mom expressed her hopes again in a recent note to me. Both of us have seen how TMS has opened up a bright new world for some people with depression. "In the future," she wrote, "TMS for autism may do something similar, allowing people to be who they are meant to be; expressing the full fragrance of their personality; being fully employed and engaged contributors in the workforce and the community; and forming warm and meaningful connections with other people."

Mind Readers

IN THE SUMMER OF 2012 I witnessed firsthand a technology that begs to be combined with TMS and other emerging stimulation techniques. Scientists are taking tentative steps toward realizing this combined power right now, and when that happens, I believe it will turn neuroscience on its ear.

We're all familiar with medical imaging. You may have gotten an MRI to look inside your lungs or to find out what's wrong with your knee. I saw the results when Alvaro's researchers did an MRI to make a 3-D model of my brain, with no hair, skin, or skull in the way. And they assured me it was accurate, right down to the whirls and folds of my cortex, where they'd placed blue and red TMS targets. What I never suspected at the time was that the state of the art in brain imaging is a quantum leap beyond that, as Dr. Marcel Just would show me a few years later at his lab at Carnegie Mellon University in Pittsburgh. He's not just creating images. He is looking at patterns of activation as we see and hear things, and by recognizing those patterns, he is reading minds. And that's just the beginning.

I was initially invited to CMU—and its neighbor the University of

Pittsburgh—by Dr. Nancy Minshew, Marcel's partner in research. We had originally met at IMFAR, where she'd asked me if I'd be willing to join a study they were starting. We took a break from the conference to talk about their work and what my role might be.

The two of them had developed a theory about how different brain areas are interconnected and how those connections differ in people with autism. They hypothesized that the connectivity between the frontal cortex and other parts of the brain was different in autistic people. More specifically, they believe the bandwidth of the data paths in and out of the frontal cortex is lower in folks like me. That's sort of like saying that autistic people have dial-up modems to connect the frontal cortex and the rear of the brain, while non-autistic people are on high-speed cable. That does not make us dumber, but it sure impairs our ability to process some things. At the same time, though, it may help some of us do other things better.

Many types of thinking, particularly social processing, depend on coordination between the front of the brain and other areas, and in autistic people, the patterns of thought we develop are altered by virtue of this poorer coordination.

"We call that 'frontal-posterior underconnectivity,'" Marcel explained.

Who wouldn't volunteer to have his head examined by someone who could utter an explanation like that? And I had another incentive to visit them. Their offices were just a few blocks from the Cathedral of Learning, the tower where my dad taught philosophy when I was five years old. That's where I learned to ride a bicycle, and I still remembered climbing the endless flights of stairs inside.

Unfortunately, I didn't have time to climb those stars when I got to town, because the researchers popped me right into a scanner. I lay inside for half an hour—far longer than I'd had to lie still back at Harvard—as their machine shot thousands of images of my brain. What they then did with that data was fascinating.

The individual images looked like slices of my brain, as if I'd been dissected and sectioned as I lay in the machine. They used a supercomputer to analyze each frame and to map the flow of water molecules along the pathways of my brain. When an MRI scanner is used this way—Marcel called the technique "diffusion weighted imaging"—the flow pattern reveals the location of the white matter tracts in the brain. They are the brain's cabling system, a complex tangle of biological wires connecting all of its different regions. As he said, "It's a little like unscrambling the connections among a bunch of water hoses in a pitch-black room by being able to see exactly where the liquid is flowing."

Later on I got to see the map of white matter tracts of my own brain—it looked like an incredibly complex and strangely colored wire-frame sculpture displayed on a computer monitor. It didn't look anything like my head, but Dr. Just assured me that the maps of other research subjects looked the same. My brain data was added to his collection, as I did my small part to help him make what will eventually become a 3-D atlas of the pathways inside the brain. One day, that atlas may become an important diagnostic tool, but today we gaze at it like medieval mariners reading a map that says, "There be dragons." We see an immensely complex three-dimensional network of connectivity in the brain, with precious little idea what the waypoints on its roads may be.

Connectivity is best imagined as the road map of your mind. You know that different parts of the brain do different things. Your visual cortex processes the data from the eyes. The motor cortex operates our arms and legs. The frontal lobes are the seat of abstract thought. All those disparate areas are tied together with a network of biological fibers, like wires between servers in a giant data center.

Brain scientists have long known of the corpus callosum—the bundle of fibers that links the left and right hemispheres of the brain. That's the network that keeps the two sides of the brain in balance, and important as that is, it's just the tip of the iceberg when it comes to brain pathways. Marcel and his researchers are analyzing scan data to render other less-known bundles of wire visible on a computer screen, to help visualize them and understand how they work. As I saw, they are getting some striking results. To explain them to me, Marcel suggested thinking of brain pathways as our highway system. He's identified a few big trunks—bundles of nerve fibers—that function like interstate highways: I-80 crossing the country and I-95 going north to south. He's also identified smaller pathways that serve as the state highways of the mind and a network of local highways, in addition to countless streets, lanes, and drives.

Our interstates make it possible to transport goods and services rapidly to wherever they are needed, and we appreciate the benefits an efficient highway system brings our economy. The highways in our brains have to be able to move information, as opposed to goods and services, very efficiently. Dr. Just's current model reveals some two hundred thousand different paths connecting the regions of the brain, and that's far from a complete picture.

He thinks that the human brain is constantly facing challenges like those solved by a navigation system in your car, but on an immeasurably more complex scale. The people whose thoughts travel the interstates of the mind more efficiently can get things done, and fast. They are the stars of mental organization. The people whose brain navigation keeps their thoughts on local streets and back roads are disadvantaged, because they can't move the data their brain requires to operate efficiently between the areas that need to be coordinated.

My friend Temple Grandin talks about this a bit in her 2013 book *The Autistic Brain.* She agrees that mental organization is key to successful independent living, and that's a well-recognized problem for autistic people like us. Dr. Just has studied both of us in his scanners. He's shown us areas where we lack paths that ordinary people have, theorizing that those differences may explain some of our areas of disability. He's also shown us places where we have more connections than the average person, which may explain our unusual skills and abilities.

Impressive as that was, I discovered they were doing a lot more. And that was the real story, as far as I was concerned. "We've learned how to read a number of emotions from brain imagery," Dr. Just told me. "And we can recognize distinctive patterns that appear in your brain in response to things that you see. We can show you pictures of a dog and a house, and the brain scans will tell us which image you are seeing at the moment."

I was stunned by his words. If that was true, the researchers were reading minds! I wondered why some government official hadn't swept in and taken over his lab. *Maybe they just don't know about it,* I thought, but I kept quiet because Dr. Just wasn't done talking. (In 2016, as this book goes to press, most of Marcel's new research is funded by defense-related government agencies.)

"One of our questions is whether patterns of brain activation are the same for everyone, or if they vary for people with neurological differences. We explored that in a group of kids with reading difficulty. We looked at the areas that light up when typical people read. Then we looked at the areas that lit up in the kids with reading difficulty and spotted some differences. When we provided one hundred hours of reading exercises that strengthened the inactive areas, the kids became better readers. The most dramatic finding was that a white matter tract whose function was deficient before our training improved after, to the same level as in kids who didn't have a reading problem."

What a brilliant piece of work, I thought. And that was where TMS came in. "We wondered if we might combine the data from our imaging with technologies like TMS to find and fix weakness in a very targeted way."

That made me think of the road maps Dr. Just was making. If he could identify roads that were less passable in people with disabilities, could TMS open them up? That was essentially what Alvaro had proposed at the start of the TMS work. Now Marcel had effectively shown us how to find those cognitive choke points. Lab tests have shown that making a neuron fire repeatedly causes it to grow more white matter around its axon and become "stronger." If TMS can help build up neurons, then the big question becomes where to stimulate. Alvaro and Lindsay had to choose targets based on research in animals, whose brains do not always mirror humans, or on general knowledge of the brain. That was a chancy approach because stimulations a few millimeters apart can have totally different effects. But Marcel's imaging techniques might provide precise individualized targeting data, and that could help create the winning combination.

Marcel agreed. "We didn't use TMS in the dyslexia study," he told me.

"But I certainly wondered if TMS might jump-start the process or super-charge the result."

I couldn't agree more. It seemed like Alvaro's team and the CMU/Pitt folks would be perfect complements to each other. If only research worked that way! But it doesn't, as I've learned in my experience reviewing grant applications. Scientists tend to come up with ideas and then approach funding agencies on their own. They often don't know what distant researchers are dreaming up unless they meet at conferences or read their published results, and there is no central coordinator.

If you think that's a major weakness in the way we fund medical research, you would be right. Most of the people who review grant applications are researchers themselves, with allegiances to their own institutions. And even if they aren't, they are bound to keep what they read confidential. If a Yale reviewer sees two complementary applications, one from Duke and another from Penn, she cannot put those people together, even if that would be better than funding either alone. Government contracting officers can do this, but it seldom happens, for a variety of administrative reasons. The rules that protect scientists' intellectual property hold everyone back in that regard.

I began to see myself as a possible bridge, because I'm not a scientist and I'm not in conflict or competition with any of the researchers. By speaking up about the promise of technological combinations I hoped I might somehow help make it happen. As smart as each scientist is, he cannot possibly keep on top of developments in every field, especially those far removed from his areas of interest.

The imaging technology being developed in the CMU lab is a good example of that. All medical researchers know what basic MRI is; radiologists have been using those systems for many years. (Marcel reminded me that Paul Lauterbur hatched the idea for the MRI scanner in a hamburger joint right next to CMU, an achievement for which he was later awarded a Nobel Prize.) Functional MRI is also fairly well known. In fMRI, the scanner makes a series of images that show the distribution of oxygen throughout the parts of the brain that are working actively, usually while the subject is performing a task.

That's enough for most physicians. Dr. Just had been in that same place a few years earlier, and he wanted more. He asked what would happen if he joined forces with CMU's computer science people and they used a supercomputer to analyze the image data. Instead of using humans to look at computer monitors, computers would run millions of incredibly complex analyses. The results of that effort are stunning in their breadth and capability.

Dr. Just uses a Siemens scanner to read activity in a subject's brain as he or she views pictures, words, or sentences on a computer monitor. When the person in the scanner sees a photo of a golden retriever, many different areas in their brain light up in a characteristic pattern. When that happens, the computer recognizes that pattern and says, "Dog." Next comes an apple, and the computer says, "Fruit." This takes place without the subject uttering a word. The scanner is simply reading the brain's activity in response to seeing different images. The computer is powerful enough and fast enough to compare that data against a database Dr. Just is building, and when a match is found, "Dog" or "Fruit" pops out.

And that's not the end of it. Computers can analyze brain scans to reveal a person's inner states of mind and possibly even intent—machines may soon predict what we will think of next! Dr. Just's system can currently recognize most major emotions, even multiple emotions that often occur together. So he can tell what we are looking at, and how we feel about it, with a startling degree of accuracy. And they have only just begun. They're getting all this information from an analysis of just 1 percent of our brain mass. A few years from now his computers may recognize several thousand words and a hundred emotions, at which point the machine could truly hold a conversation with your mind and you wouldn't have to say a word.

The possibilities for treating disorders of the mind are staggering. Comparing the brain activity of an ordinary person with that of someone with a disability, we can see with unprecedented specificity how the disability affects the working of the mind. That may lead us to ways of normalizing or changing those brains and fixing disorders at an elemental level.

When I suggested that to Dr. Minshew, she agreed. "Imagine a future where the evaluation for people with concerns about their mind and brain involves a short interview to define their needs and goals. We'd follow that with brain imaging and a blood draw for genetic sequencing. Those test results will yield a precise diagnosis, identification of the specific cognitive, brain, and genetic mechanisms causing the person's concerns, and a treatment plan for the mind and brain that is specific to that person.

"Imagine that," she continued. "That is what scientists and clinicians are striving together to attain. Dr. Just's neurocognitive research is only one example that shows that such a future is feasible. The potential life improvement from combining rTMS [repetitive TMS]* with emerging cognitive rehabilitation treatments, informed by the latest genetic advances, is huge."

We are truly on the brink of a new era for treatment of the mind.

---

* In this book I have used the term "TMS" to describe all the stimulation techniques I experienced. To be more specific, therapeutic TMS uses multiple pulses and is also called rTMS or burst TMS. Single-pulse TMS is used more for measurement rather than treatment, as when Lindsay applied single pulses to find the power level necessary to make my fingers twitch.

A Death in the Family

AS A YOUNG MAN, when I encountered death on the highways or violence in the nightclubs where I worked, my autistic oblivion provided me with a kind of psychic protection. *Sort of like Novocain at the dentist,* I told myself when I figured it out. As I got older and had more life experiences, the anesthetic wore off and I felt more empathy for those around me. I'd always had feelings, of course, but autism had prevented them from being triggered in many situations; however, once I'd scraped my own leg a few dozen times I began to get a sense of what someone else might be feeling when that happened to her. I might not have felt the same way others did, but my responses still moved more into the socially acceptable range. Good as that was in many ways, my autistic protection began to dissipate. Then I met the TMS scientists, and they worked their magic on me. That was the end of my detachment from death.

That's partly why I felt such fear and dread when I got a text message from my first wife, Little Bear, in the summer of 2013. "I may need you to take me to the hospital," she'd said. I didn't see the message right away, and by the time I could answer she had already gotten herself there. I followed as quickly as I could.

When I arrived at Cooley Dickinson Hospital she was resting on a bed in the emergency room, waiting for news. Her blood pressure had collapsed. She felt weak, and her heart was racing. Their first thought was internal bleeding, but that was ruled out a few hours later. That left anemia, or something worse. Cubby came down from Vermont, and Maripat arrived a few minutes later, but none of us knew what to make of this news.

Something about the situation struck me as terribly ominous. Nine years earlier, I'd walked into a hospital room after my father fell down. He was almost seventy, and he'd been in and out of hospitals a number of times in the previous two years. That time didn't seem much different from all the others. But somehow I knew it was. As soon as I was out of his sight I'd started crying because I knew he was going to die.

Now I had that same feeling looking at Little Bear. She wasn't old, she wasn't crippled, and there had been nothing to suggest it in her words. Still I was sickened when I heard her voice. Something inside me said, *She's not going to make it.* I knew it with absolute certainty, and I felt like throwing up. Even now I cannot say why I had that feeling. We had been separated for many years but I always loved her, and she was the mother of my son. If she went, a piece of me would die with her.

And I wasn't the only one who started worrying. She was very close to Cubby, and Maripat and she referred to each other jokingly as "sister wives." Mary taught Maripat to shoot handguns with her at the Smith & Wesson range, and though a few of their escapades made me twitch my ears a bit, the way they brought the family together in the three years they'd known each other was truly remarkable.

And now this. The blood transfusions restored her energy, but they did not fix the problem. Even worse, we learned about a new problem: immunity. The body protects itself from disease with white blood cells, and transfusions don't replace white cells. The only way to get them would be for her body to make them. Her white counts weren't zero, but they were very low.

"That's a real danger," her doctor said. "People with low white counts are vulnerable to any infection at all—things a healthy person would shrug off without even noticing. There must be something wrong with her bone marrow."

That's what anemia and leukemia are—diseases of the marrow. One's a cancer and the other is, well, just unknown. They ran all kinds of tests on her blood, tested her for known cancers, and found nothing. We wondered what was wrong, knowing deep down that whatever it was, it wasn't going to be good. The hematologist at Cooley Dickinson wasn't sure what to do with her. He suggested transferring her to the specialists at Mass General, two hours away in Boston.

Mass General is another of the Harvard teaching hospitals, and it's one of the best in the country; if you have a serious problem there's no better place to be. Mary spent a month out there, but the best doctors in the country couldn't find an explanation. The facilities were beautiful. She was on a top floor, in a private room, with a view of the city. And unfortunately, the choice private room wasn't because she was a pampered celebrity. She was in a room of her own because her immune system was failing. That's what they told us. Anyone who went to see her had to disinfect at the door and put on a mask. The air in her room was specially filtered. They took every step they knew to prevent her from getting an infection, because anything can kill you if your immune system isn't working.

Eventually her white counts went up and they said she could go home. All of us pitched in to clean her house as spotlessly as we could in time for her return. A few weeks later, after coming home from the hospital, Little Bear went back to work. She had math and science classes to teach at Salter College and work to do in the hacker space on guitars. She took an order from Ace Frehley for a new light guitar and the repair of his old one.

Every couple of months, Maripat and I took turns driving her to Boston for transfusions of blood at Mass General. We'd sit in the waiting room and read as she reclined in an easy chair and blood dripped slowly from the bags into her veins. She'd be weak and pale when we went in and flush and zippy when we left. The effects of the new blood were immediate, but transfusions can't go on forever. Just as transplanted organs are rejected, the body rebels against blood that isn't its own. The more times you do it, the more your transfusion options narrow and the harder it can get. Each time, we hoped the next transfusion would be the last, but in fact each one lasted less time than the one before, and as summer turned to fall, she had to call even sooner to get transfused. But the changes in her body made matching blood harder to find, and by wintertime we had to wait ten long days for her transfusion as they hunted all over for a match she wouldn't reject.

There were other procedures too. She and Maripat went through most of those together. They did three bone marrow biopsies, and none of them showed a thing. Back at the Northampton hospital they monitored her blood counts weekly. Some weeks she was up, others she was down. She was making new blood cells, but not quite enough. The transfused blood was keeping her going, but everyone knew it wouldn't last. Then, in mid-January, things changed. Surprisingly, the change was for the better. Or so it seemed.

Little Bear was so excited that she shared the news with us and her five-hundred-some friends online. She wrote, "I had the possibility of REALLY good news today. I saw my local hematologist after 'the usual' blood tests. I'm having a bad asthma flare and am on my second course of oral steroids since New Year's. Suddenly, all of my blood levels are normal or near normal, after seven months of extremely low, not good levels. The local doctor said I may NOT have aplastic anemia, and all of my blood problems might be due to an autoimmune disorder. That would be a LOT better because aplastic anemia does not have a terrific life expectancy after diagnosis. I'll see my doc at MGH in two weeks and she will have presented the current results to the team. Wish me luck!"

It sure sounded hopeful. *Maybe it's a miracle,* I told myself. Keeping quiet about my earlier premonition, I hoped with all my heart I was wrong. That weekend, she went off to the Arisia science fiction convention and had a chance to connect with a bunch of her old friends from the UMass Science Fiction Society. That was where we used to hang out in 1980, when we first got back together after high school. She had to wear a gauze medical mask to the convention, because of the autoimmune warning, and Maripat's daughter decorated it with a cat face for the occasion.

The following week she was out and about again. Salter College had its graduation ceremony and she was proud to don the robes she'd worn at her own doctoral ceremony nine years previously. In that time we had learned she too was autistic, and that knowledge went a long way toward explaining why education had been such a long and difficult road for her. She was just beginning to come out of her shell and talk about that, particularly with younger autistic women.

Two weeks later, the steroids were over and done with but the blood counts remained normal. Things were looking good. If only it had lasted. . . .

Somehow, everything changed in the space of a week. Little Bear celebrated February 5 in the local ER, after a bad blood test the night before. She'd gone to the doctor for a routine blood draw, but her readings were so far out of whack that he called her that evening. The high white counts got them thinking internal infection and immediate hospitalization. It didn't make sense. "I've never felt this good," she said, "while having a doctor tell me to call the ambulance." She made a date at Mass General and got ready to drive there in the morning.

Maripat took her to Boston early the next day. They checked in, and Little Bear had more blood drawn. The results were back in an hour. This time her blood was full of blast cells. Alarmed by that, the doctors did a bone marrow test and came back with a diagnosis of acute lymphoid leukemia. Cancer had reared its head, and in an aggressive and virulent form. "I'm just glad to have an answer," she told us, but we knew she was terrified. The doctors must have shared her concern, because they decided to start the chemo in the morning. Maripat stayed on the sofa by the window and they faced the first night together.

They would be together for quite a while. Hospitals have always made me very uncomfortable, and I could not bring myself to stay there. My son seemed to share my reluctance, but Maripat was strong and present. Little Bear depended on her most of all.

She was surrounded by technology—hooked up to three computer-controlled IV lines, monitored on two TV screens, and talking to the world on Facebook. Day one included a spinal tap and chemo into the spine, because this kind of leukemia likes to linger in the central nervous system. When it's aggressive like hers they want to nuke it immediately. This first round would be the longest—about three weeks—and the toughest, with the highest doses. Once she got through that, she was scheduled for seven more rounds—each lasting a week, at one-month intervals.

That was when the real demon reared its head. "I've had this stupid little cough for about ten months. They took a spit sample on the second day and found Aspergillus fungus. It's time to start nuking it before I get into trouble being immunocompromised. Better a dead fungus than live mold eating my alveoli."

She sounded enthusiastically pugnacious, though every complication worried me more. But I couldn't show my fear. "You can't know what will happen," Maripat told me. So I kept quiet as best I could. I reflected on our lives, together and apart. Little Bear and I had first met forty-three years before, in seventh grade at Amherst Junior High School.

Both of us are so lucky Maripat came into our lives, I realized. I couldn't imagine what we'd have done without her. Cubby and I get terrified, and we have to run away. Even with all my TMS and personal growth I still can't handle hours in the hospital. I guess Maripat gets scared too, but she has the inner courage and strength to stay when we can't.

That night, Little Bear had her first weird dream. She had to get to school but the interstate was a bit backed up, which was why riding her tricycle in the middle lane was no problem. None of us knew what the dream meant, but I listened to every word she said now. The next day Maripat called me at four o'clock, while I was still at work. "They have to do an emergency operation. You should come out here right now."

Earlier that day, the doctors had discovered that the mold had spread to both lungs, and they said the outcome when that happened was "usually not good." Now they wanted to cut through the roof of her mouth, remove the palate, and cut the infected tissue out of her sinuses. It sounded horrible. I asked to talk to Mary.

She got on the phone and told me in a clear voice that she'd decided to do the surgery. "It's the only chance," she said.

I told her I loved her and we would all pray for her safety. I told her I'd drive out and be there when she woke up. While I was on my way, I got another call. More doctors had come in, and they'd reconsidered the operation. They decided to monitor her progress with the medicine and wait a few days. She was too weak for surgery.

When I got to the hospital she looked terribly weak. Maripat was petting her shaved head, and I walked over and petted her too. The moment I touched her, she purred and smiled. I knew it soothed her. Maripat sat down, and I petted Mary softly till she fell asleep.

Before I arrived Mary had told Maripat that she finally realized I'd never meant to hurt her, that she knew I'd always loved her, and that she now realized I'd never been her enemy, even when we were getting divorced. It came as a shock to hear that she was still thinking all those things. It had been fifteen years since our separation.

I felt so sad. She was visibly weaker and I knew I was witnessing the last days of her life, just like when my dad died. I resolved not to let her see me cry. Maripat said that we needed her to keep up hope, because there was a chance she could beat the infection. *The odds are long, but it's possible.*

Maripat was worried because Mary hadn't eaten all day. When people are dying they often stop eating. The nurses told me that when my dad was sick.

Days passed, and family came and went. What seemed like a hundred friends came to say goodbye. Maripat, Mary's sister Karen, and Cubby had spent every night of the past weeks beside her. My stepson Julian went above and beyond too, keeping our home together while Maripat was in Boston, cleaning up and caring for Mary's house, and spending time with her in the hospital. Now they were all home getting some much needed rest. Finally it was just the two of us, alone for one more night. I'd visited her a number of times during this sickness but this was the first night I'd stayed.

She was very weak and barely able to sit up in bed. She couldn't say more than a few words, and you had to listen closely to catch what she said. We both knew the end was near, and she was scared. Being alone terrified her, as did falling asleep. She desperately needed to relax, but the idea that she might never wake up kept her in a state of near panic.

People talk about dignity in death, but there is no such thing. Grave illness knocks you down and narrows your vision. When it's a strain to draw a deep enough breath to remain conscious you can't think of anything else.

It's comforting to have someone beside you, holding your hand. My own scary experiences in the hospital had shown me that, when I'd had a combination of asthma, flu, and pneumonia, and I saw that with Little Bear. Holding her hand or gently stroking her head had an immediate calming effect. It may not have cured the disease, but it dissipated the fear for a little while and brought some healing relaxation.

She couldn't say much, so I told her about my day at the car company. Then we talked about something new—a class I was scheduled to teach ten days later at the College of William & Mary in Williamsburg, Virginia. Little Bear had started teaching while she was getting her doctorate at the University of Massachusetts. Now I was getting invited to teach college courses, too, both locally and far away. The William & Mary connection was particularly interesting because one of my ancestors had taught there at its founding, and I'd always been fascinated by history—particularly when I felt a personal connection to it. "The Jamestown historic site is just five miles away," I told her. "I'll go there and send you pictures of what those Virginia archaeologists are up to." She smiled and nodded her head, remembering her own grad school days doing archeology at Historic Deerfield, Massachusetts.

Eventually I ran out of things to say, and we settled in for the night. She was in her hospital bed, and I had a cot beside her in the room. It was hard to fall asleep, and I woke up every time she coughed or moaned. "My neck hurts," she said, and I padded over and rearranged pillows. She was in a soft bed with lots of pillows, but there just didn't seem to be any position in which she was comfortable. When I fell asleep she started coughing and choking. "Aaaaaah," she moaned. I said, "Woof," and she settled down, comforted that I was still there. We passed the night that way, in a sort of companionable sleeplessness.

I had to leave the following morning, but Cubby and Maripat were at the hospital to take care of Mary. The first thing I did when I got to Virginia was drive straight to Jamestown and text pictures of everything I could see to Cubby. He showed them to his mom, one by one, and she smiled. Things were stable for another five days, but then they took a turn for the worse. Cubby called me at nine-thirty on Monday morning, his voice shaking as he tried to stay calm. He sounded very logical—as he always does—but I knew how much he was hurting inside. Seeing his mother struggling and in pain these past weeks was surely the hardest thing any of us had gone through. "It's pretty much over," he sobbed. "She kind of stopped breathing, then she twitched, and then opened her eyes and rolled them up. And now she's not responsive at all." He had stayed at her side as her breathing got shallower and shallower. *At least he is there beside her,* I thought. I cannot imagine any greater gift he could bring her as she left this world. "I think she knows I'm here," he continued. Then he started to cry again. Four hundred miles south, in my hotel room in Virginia, I cried too. Writing this today, a year later, tears run down my face again.

At twelve minutes past two, Cubby called again, still sobbing. "She's gone. . . . It was peaceful. . . . I was reading, holding her hand. . . . She just stopped breathing. No choking or anything. I closed her eyes and waited ten minutes to see if she would come back. Then I called you. The nurse was there, and she listened for her heart. I guess I have to go back and cover her up."

Three days later, she was buried beside her brothers at Granby Cemetery. She was fifty-six years old.

Back in the Groove

SPRING CAME, bringing a measure of light after a dark and sad time. The summer gave us a few months to recover, and then it was fall—the season for harvest festivals and fairs. Six and a half years had passed since the night TMS transported me out of my car and backstage to a Boston soul concert. Vivid as that memory was, it was just a hallucination. Now September was here, and I found myself waiting for another show to begin, one that was real and happening all around me.

The lights had already dimmed, and the audience was stirring in their seats. Across the yard in the concession area, bartenders were pouring beer as fast as they could; they knew they were on borrowed time. Up on the elevated stage, twenty feet back in the dark, Gary Rossington and Rickey Medlocke of Lynyrd Skynyrd tuned their guitars. Maripat was settling into her seat out in the audience, beside my buddy John Juliano and his wife. John produces the concerts and events for the Eastern States Exposition—otherwise known as the Big E—and I come to West Springfield every year to photograph the shows and the spectacles. It's always good to have friends in management!

As public as my car business and writing are, photography sometimes feels like an activity I do in secret. I've been photographing performers at rock concerts, circuses, and fairs for fifteen years, and artists and venues license my images, but few viewers know they are mine. That's the funny thing about photography. You pick up a book and see the author's name on the cover, but hardly anyone knows who takes the photos on billboards or in magazines, even if millions see them.

In the time I'd been taking photographs, I'd developed what I thought was a recognizable style. If you look at images I shot in 2007, you will see similarities to pictures I took in 2005, or 2002. I thought of my photographic style the way I viewed the grille on a car—it might change and evolve over the years, but if you recognized a brand of car by its grille in 1968, you'd know the same grille in 2008.

That was true for me until the summer of the TMS study, when my style changed suddenly and radically. The colors in my photos suddenly got brighter, and the compositions got simpler. And viewers liked them better.

It took a while for me to connect the alteration to TMS. I didn't actually notice that a change had occurred until the following fall, when I switched my photo catalog software, and a scan through my new image library in chronological order showed an explosion of color after the summer of 2008, right after the TMS experiences.

At first I wondered if I'd changed cameras, or processes. But I quickly realized that wasn't the case. Cameras are usually used to render color the way we see it, which is not bright and oversaturated. But my images had become colored to the point of looking artistic because I set the camera that way. It must have been a conscious decision, though I can't remember thinking about it at the time. Like the musicians in *Spinal Tap*, I'd started turning up all the dials on my camera as far as they'd go. Suddenly, I remembered something my son had said after a TMS session that summer. "Dad, the colors around me are all brighter, and I can see more shades." I hadn't known what he meant at the time, but now I realized the same thing must have happened to me. And it hasn't faded away. Instead, it's become an ingrained part of my style.

My pictures used to be like the soup containers on the grocery store shelves—accurately colored, properly composed, and boring. Now they are like Warhol soup cans—brighter and further from reality but more real to the viewers. They garner far more positive comments, from both performers and the public. At the same time I started moving in closer and framing my shots tighter. The result is a picture with one central element and fewer distractions. It's less realistic than real life, which is full of distractions, but people see it as more real.

Before TMS, I would shoot properly focused and framed head-to-waist torsos of musicians playing guitar onstage. And those pictures were okay; the musicians praised them and the public seemed to like them too. But as a comparison, in 2008 I shot Barry Goudreau, guitarist of the band Boston, and in one frame all you see is his hand, the face of the guitar, and the strings. Abstract as that is, it's ringing with life and color. The close-cropped image is so much more powerful, even though you can't see who's playing. I'd never done work like that before TMS. Now, it's all I do.

So there I was, right up front with camera in hand. It was unseasonably warm for late September in Massachusetts, and the fairgrounds were so crowded you could barely move. Over a million and a half people make the pilgrimage to West Springfield for the Exposition over its seventeen-day run, and I was there every moment I could break free from the garage and my other responsibilities, walking around, seeing the sights, and taking pictures. There's no knowing when the stars will align and I'll shoot a winner.

Taking a moment, I reflected on how my life had changed. Sometimes I felt as if I had been turned upside down and shaken, even though any direct effects of TMS had long since dissipated. In some ways it seems like I'm back to being oblivious, but I still feel significantly changed when compared to my pre-TMS self, and the response from my friends confirms it. They've sure said that on the fairgrounds.

Today I'm a lot more ready to accept that other people may have abilities very different from my own. Now I realize that most of the people around me can look in another person's eyes and see his feelings. Even if

I can no longer do that myself, the knowledge that I did it once—and that others are doing it all around me—is enough to step up my ability to interact with others. And it's a change for the better. I see that every day, but it's most visible when I have to have lots of quick encounters with strangers. That's exactly what happens on nights like this, when I'm in a crowded arena shooting images of rock stars, circus acrobats, carnival operators, and their fans. My night at the fairgrounds might include a hundred quick hellos and dozens of conversations with the people in my photos, fair workers, and managers of all stripes. Now that I've been magnetized, those things just go smoother and better.

As we waited for the show to start I talked to Mark Murray and Don Treeger, two other photographers who'd been hired to capture this spectacle. Mark and Don are long-time newspapermen, masters at the art of rendering today's scenes for tomorrow's front page. Years before, I'd tried and failed to engage people like them in conversation. Now I do it easily.

Guys like Mark or Don used to make me feel uneasy; they were intimidating. After all, they are pro photographers and I am just an enthusiast. Sometimes I felt like they were adults and I was still a teenager, even though that's decades in the past now. Today I realize they are just guys like me, there to do a job. It's hard to put my finger on exactly what's changed. All I know is, I can now be around them without feeling anxious, inferior, or threatened. If you've never felt that way it may be hard to imagine, but for me it's a big deal.

My fair-time acquaintances had picked up on the TMS-induced changes right away when they first appeared. They form memories of me but then they don't see me for another twelve months, so it was a surprise for them when I seemed different and better, with no apparent cause. You expect kids to grow up and be more socially skilled but you don't expect those changes in middle-aged adults.

One person who commented on my development was Charley Van Buskirk. He's been the Big E's master of ceremonies for as long as I've been alive. Well, maybe not that long, but it feels that way. He knows everyone, and he's well loved and respected. He's also been around me enough to see how I was, and what I became. "You were always a very talented photographer," he told me recently. "But you were also difficult, abrasive, and socially inept. I actually avoided you. Then, a few years ago, you changed, and it was dramatic, to say the least. Now you're a sociable and likeable person that I seek out."

Gerard Kiernan said something even more surprising when I ran into him. Gerard is in charge of all the buildings and fairgrounds, and he's the guy I turn to to get things moved or rearranged for a photo. Busy as he is, I figured I'd be the last person he'd want to see. "Not so," he told me. "You've become one of the most insightful people I know and I look forward to talking with you." The idea that he'd say that about me—a middle-aged autistic guy formerly known for his social oblivion—was remarkable.

Embarrassing as it was to hear how I used to be before TMS, it's also kind of hopeful or inspiring to imagine that I could become so different thanks to a little change in emotional processing. And the change made me friends all over the fairgrounds. I was respected before, but now people liked me. What a cool thing to experience, especially for an autistic guy who grew up alone! Ten years ago, people just ignored me as I took my pictures. Now I sometimes feel I hardly get a chance to use my camera with all the folks who walk up and talk to me.

I waited excitedly with the rest of the massive crowd for Lynyrd Skynyrd to start their set. I'd come of age hearing songs like "Sweet Home Alabama," "Gimme Three Steps," and "Saturday Night Special," and it was a thrill to hear them play again, live. Live shows are the best, because they are music at its most raw and real. It was actually kind of a freak thing that they were here at all. ZZ Top had been scheduled to play this show, but they'd canceled with two weeks' notice when their bass player broke his hip. Like most people, I figured the Big E stage would be dark tonight, and I was happy to hear Skynyrd would be here in their place. Now they were onstage, and we were almost ready to start.

Thirty years ago I'd stood on similar stages as a sound engineer. Today

I was a photographer, and another sound engineer stood in my former place by the monitor console. He was up there where I used to be, waiting for the show to begin.

There was a stirring behind me and I realized it was showtime. Charley stepped up to the microphone and announced the band in his trademark drawl. He was barely clear of the spotlight when the first guitar notes rang out. *This is gonna be a loud one,* I realized, as the bass hit me hard enough to rustle the hairs on my arms. I raised my camera and began to shoot.

They call the area I was in "the pit" because it stands like a moat between the swarming crowd and the musicians onstage. To my right, the stage stood six feet high, and I had to stretch up to see the floor. To my left, a picket fence was all that kept the audience at bay. It was already bulging inward as fans pressed up against the rails and held cellphone cameras at arm's length like offerings for the band.

I shared the five-foot-wide space with Don and Mark, three more photographers, two video camera operators, and half a dozen solidly built security men. If it weren't for them we'd surely have been trampled. As it was, we stepped over one another's gear and cables carefully as we moved from one side of the stage to the other in our quest for that front-page shot.

When I had a moment to pause and reflect, I recalled those long-ago nights in Boston when I built electronic systems and helped produce the shows my equipment played in. Back then I had been an isolated geek building electronic devices with little connection to the people who used them. Now I was part of a community. There was no way I'd have ventured into the pit in the seventies. The people outside the fence seemed like wild animals, waiting to tear me apart. It could be suicidal to go near them, and I didn't. Why was I different now? Partly, I'm older. But the biggest change—the thing that set me free to go shoulder to shoulder with a mad crowd and shoot the best images of my life—came to me after TMS. That's what gave me the understanding and confidence to go among people, secure in the knowledge that I was part of a community, and I was safe.

As I moved smoothly between the security guys and camera crew around the stage, I realized that TMS had helped me join the community of man in a way I never had before. A few hours earlier I'd been walking the fairgrounds with my old friend Gene Cassidy, talking to people we passed. "You're like a politician," I said to him after his tenth or twentieth handshake, and he laughed and agreed. But I was right there beside him, something no one would have predicted from my behavior thirty years ago.

When we met, Gene was a young accountant and I was starting my car business. Now we were middle-aged, and he was the president of the fair and I was, well, whatever I became. We'd shared many confidences over the years, including the time I'd learned about my own Asperger's, fifteen years before. Gene reminded me of that recently.

"I still remember the call," he told me. "I picked up the phone and said hello. You didn't even greet me. You just said, 'Do you think I'm weird?'

"I didn't know what to say, and you repeated your question. All I could say was 'No—I think you're John.' What kind of a question is that? Then you told me that some psychiatrist or something of the sort had just suggested that you showed traits of this disorder called Asperger's. You said it was some kind of autism.

"You said you wished you had known that when you were seventeen, and I could tell you were really affected by what you were talking to me about. Very affected, and sort of sad. Hearing you that way made me feel bad. It also made me mad, because I thought whoever the shrink was, he had some nerve saying that to you. I didn't understand it, so it seemed hurtful."

I've gone from being a machine person who interfaced with humans when he had to to a people person who understands technology. And that's just one part of my life that's come full circle. I left music because I wanted the stability of a "real job." Now I was back on the stage, shooting photos because I had a drive to be creative that overpowered my sense of duty to the "real job."

The band started its second song with all of us down below shooting away. As the notes rang out, my concentration sharpened, and my sense of who and what was around me disappeared. I was purely in the moment, living the scene in my viewfinder and pressing the shutter when the feeling seemed right.

I tried to recall all the performers I had photographed over the years, and with a start, I realized I'd photographed and interacted with more musicians since doing TMS than I had in the decades since I'd left the music profession.

Not only did I interact with more people, I made more successful connections. That was obvious, walking around earlier with Gene. With the clarity of hindsight, I can understand what happened. My technical achievements and commercial success had made me respectable before TMS. But respect and friendship are two different things. While many people may have respected me, comparatively few wanted to be my friend. Now everywhere I went I felt welcome, walking round the fairgrounds.

Years before, being respected had been my defense against being dismissed, criticized, or even bullied. Now, seeing the alternative, I realize that being a friend is immeasurably better.

The lights onstage changed color, and the spotlights swung to Peter Keys, hidden way in the back on stage right. With his leather jacket and tattoos, he could have been mistaken for an outlaw biker working backstage security. Then he sat down at the keyboard and hammered out the opening chords of "Free Bird." The audience roared and I looked around with a start, realizing the concert was almost over. I'd been concentrating so hard I'd spent the last hour on autopilot, and the frame counters on my two cameras showed I'd taken 625 images. A few would turn out to be some of my best shots ever.

I'd started out in the pit in front of the stage, but the band only allowed photography up there for the first three songs. Afterward, I'd spent the show walking the grounds, looking for interesting compositions with the audience and the stage in the distance. Now I stood one hundred yards out in the crowd, surrounded by a massive sea of people—a place I would not have been caught dead in twenty years ago. Yet I was at peace.

I may not have recognized anyone around me in the dark, but everything was all right. Even with the roar of the crowd and the music, I felt secure knowing my wife, John, Gene, and all the others were out there somewhere. I smiled to myself and headed for the gate.

Postscript: The Future

AFTER THE RELEASE of *Look Me in the Eye,* I received invitations to a number of colleges, schools, and even elementary and preschool programs. One place I visited was the Ivymount School in Rockville, Maryland. I was invited there by Lisa Greenman, the mother of one of their students. She introduced me to Monica Adler Werner, who heads Ivymount's Model Asperger Program, and that spawned a series of collaborations that lasts to this day. Monica showed me a curriculum they were developing—called Unstuck and On Target—that was aimed at helping autistic kids get organized and moving. I was so impressed by it that I wrote a foreword for their workbook, and I became an informal advisor to the program. As much as I believe in the power of TMS, I also see the value of talk therapies that help people get their lives together.

Monica in turn introduced me to Karin Wulf, the mother of another Ivymount student and a professor of history at the College of William & Mary. Karin invited me to come to their campus, saying, "I want to make our college a friendly place for students on the spectrum." We talked about the new neurodiversity and autism rights movement that was beginning to gain traction, and Karin was ultimately instrumental in my appointment as William & Mary's Neurodiversity Scholar in Residence. Together we are developing the first neurodiversity program at a major American university, and it's a very cool thing to be part of that process.

As I've mentioned, when it comes to science and autism research strategy, I'm still not an expert, but I'm learning as fast as I can. At age fifty-eight, I'm not likely to become a researcher myself, but I'm proud to be part of advisory boards that shape research that's going on right now. I've just started to advise a group at Yale that's part of an NIH biomarkers consortium. A few years ago, the secretary of Health and Human Services appointed me to the Interagency Autism Coordinating Committee, the top-level autism strategy group for the U.S. government. It's been a great honor to help develop our government's strategic plan for autism research. I also serve on the steering committee for the International Classification of Functioning, Disability and Health (ICF) Core Sets for Autism initiative for the World Health Organization. Just reciting that list is enough to make my head spin. *How did I ever make the leap from being a car mechanic in New England to doing those jobs on an international stage?* If I knew the answer to that, I could package it and sell it for millions of dollars. There are probably many factors, but the one I think of first is TMS, for the insights it gave me, and for the confidence it gave me in myself. In that sense, TMS was like a teacher. And the scientists who answered my questions were teachers too. For me, the combination was very powerful. I look forward to the day when many others will find a similar benefit.

Meanwhile, as exciting and challenging as all that is, I still come home to my cars and to the business I founded to service them. Toy cars were my best comfort as a boy, when I used to put my fire truck up on blocks for service. Later, working on real cars gave me a way to make a living and achieve independence. Whenever I failed to connect with people, my machines were always there. Whatever else I do, they will always be there, waiting. And now there's an exciting new chapter to that part of my life. We're doing more high-end restoration work, and we're turning out some very fine pieces of automotive art. I wonder where that will lead us, and what we will work on next.

Today I split my time between the car company and autism advocacy and teaching. I speak out for the rights of autistic people and our need for community. I speak out for science, because it represents the path to new knowledge, and that's something we need desperately in the autism world. We have so many challenges, and so few answers.

One tool that can provide those answers is TMS. I know that because I've felt its power firsthand. Everyone dreams about greener grass on the other side of that fence; TMS helped me go there and taste it. And though there were some downsides to the TMS—mostly related to the bad feelings I was unexpectedly sensitized to and the collapse of my marriage—I now see that's the reality of the world. Yet we still have a long way to go toward understanding its potential. As Lindsay has told me repeatedly, "The last thing I want to read is that someone decided to make a home brain stimulator and fried himself because he read scientists were using one-milliamp currents and he decided ten would be better. People have to understand it's not that simple." I promised her that I would do my best in this book to share both the promise and the risk and complexity of these new technologies.

And I hope that I have.

Throughout history medicine has strived to fix what was broken and cure what was diseased and by doing so restore us to health. In recent decades that focus has begun to change. We've learned how to develop our bodies from merely functional to exceptional. Now, neuroscientists are showing us a similar path for our minds. Researchers like Dr. Just are creating tools that can reveal and explain brain function with remarkable specificity. At the same time, scientists like Dr. Pascual-Leone are developing new ways to reach inside the mind and effect change. When those technologies converge we will see a revolution in treatment of the mind, perhaps even the emergence of a whole new discipline: cognitive exercise and mental fitness.

Autism researchers are hard at work searching for biomarkers—biological parameters that predict or diagnose autism, particularly in infants. Dr. Just, Dr. Pascual-Leone, and Dr. Oberman have each published studies describing their achievements in this area. Dr. Just measured the brain's response to imagining emotions like humiliation. Drs. Pascual-Leone and Oberman measured the way the brain changed in response to bursts of TMS energy. Both techniques separated autistic from typical subjects with a high degree of accuracy.

There's no doubt that this is important work, all of it. Identifying autism through precise neurological measurement rather than behavioral observation is a great thing, because it reduces the possibility of error. And it increases the odds that we will remediate its disability successfully, because an understanding of its foundations will be key to changing its effect on us. The earlier we do this, some say, the better. Yet all the scientists who appear in this story have raised real concerns about changing brains in the absence of fuller understanding. They've made great progress, but much more remains to be discovered. And as exciting as those developments are, I worry that they may become steps toward a kind of neuro-homogenization. When we develop precise means of identifying neurological difference, and we pair that with targeted interventions to resolve the difference, what do we get?

In the example of Dr. Just's dyslexia study, we got kids who can now read better, thanks to science. No one would question the benefit of that. The ethical problem appears when we begin testing people for differences and then applying corrective therapies before problems have become apparent.

The thing is, not all differences lead to disability. Some lead to exceptionality. And we don't necessarily know enough to tell one from the other. Yet we are on the verge of acting on that incomplete knowledge right now in the area of autism.

The problem comes when we presume all difference is disability, and it's not. There's a growing body of evidence that some of the world's smartest and most creative people have traits of autism or other neurological differences. Just seeing things in a unique way can help autistic people solve problems that baffle our typical peers. Most of the exceptional things I've done in my life were facilitated by my being autistic. What if all that had been wiped away by early intervention?

Neurostimulation offers the possibility of rewiring our brains to fix previously unfixable disabilities. In theory, early childhood treatments might allow kids who would otherwise grow up disabled the chance to grow up without disability. Some ethicists fear that early intervention could lead to treatment of people who don't need fixing and the possible blunting of their strengths. As Alvaro showed me, treating someone unnecessarily can create problems of its own. An alternative would be to wait until kids are older and disability becomes visible, and then use treatments to remediate it. Yet some scientists fear that this approach will be too late for an optimal outcome. Children have more brain plasticity than adults, which might mean that TMS would be more effective if done earlier. But it might also mean changing things we don't intend to, because the developing brain is so susceptible to change.

In my opinion, we need to be very careful about anticipating future disability in very young kids and then applying treatments. Pediatricians know that many developmental problems fix themselves, and normalizing differences should not be a goal in and of itself. Many of humanity's greatest leaps have come from people who are different. And current tests that predict someone will grow up autistic can't tell you whether the person will be crippled as an adult, an eccentric genius, or even both. Although TMS has had great benefits for me, I worry about changing kids' brains in hopes of eliminating disability and snuffing out the fire of creative difference that moves the world forward.

Alvaro and I talked about this on several occasions. One of our questions was why the oldest people in the TMS study seemed to feel the most changed. "Maybe being older helps, because you have a bigger base of life experience to draw on when trying to integrate the effects of TMS," he suggested. I think that's quite possibly correct, and if so, we need to be very cautious about stimulating kids who may not be ready to seize what their minds are offered. Perhaps Nick is an example of that. While both he and I experienced temporary cognitive enhancements, only I was able to build a lasting change from them. Cognitive therapy—Dr. Minshew's area of expertise—may have helped. All we can say right now is that it's one more example of why we need to learn more.

Another thing we need to consider is the attitude young people may bring to treatment. Right now, for example, Nick says he has no interest in future TMS treatments. My inclination would be to respect that. But there are parents who feel quite strongly otherwise. Whose decision should that be? I always wanted friends and acceptance, but I never wanted my differences erased in favor of becoming some kind of smooth-talking robot. There's also the temporary nature of the change to consider. Some people will feel like me—they can change their lives for the better based on temporary experiences. Others will see the arrival and subsequent fading of TMS's gifts as a cruel joke, one they want no part of.

With issues like these on the table, it's clear we need cognitive therapists in the loop, right beside the technicians and scientists. They weren't a part of my experiences until the very end, when I connected with Dr. Minshew. That's not to say I was unsupported—Lindsay, Alvaro, and the others were all there for me. And in all fairness, it's been experiences like mine that have shown the need for therapist backup.

Another ethical quandary we will face is the potential for neuro-enhancement. The Osborne article in *The New York Times*—the 2003 story I'd read before doing TMS myself—outlined this possibility, and as brain stimulation technology develops and becomes more widely known, more people may seek out off-label brain stimulation in an effort to "improve" their brains. At some point, a line will be crossed. It's inevitable. In sports, it's fine to exercise and take vitamins. But swapping out your blood and ingesting steroids to bulk up muscles is forbidden, and dangerous. Yet it's frequently done. Where will we draw those lines when it comes to the mind? We hope that brain stimulation is safe when administered by experts, but an overdose of stimulation—or stimulation of the wrong places—could be just as dangerous as an overdose of drugs. And where does it end? Will high schoolers of 2026 be given TMS to improve their SAT scores? Will we end up with a black market in brain stimulation? Or will TMS for brain enhancement be banned, like erythropoietin and steroids?

High-tech brain imaging like Dr. Just's is approaching its own ethical decision points. It's one thing to study the brain activity of people with a recognized disability and compare that activity to nondisabled peers.

Using that knowledge to guide interventions that build up the dysfunctional areas seems like a great thing, and Dr. Just showed that it's possible with his dyslexia study.

But what if he flipped the study upside down? What if he had studied exceptional readers and then developed exercises to build up typical readers to that level? All of us want to be the best we can be, and I have little doubt that people would volunteer to take part in a study like that. We all want to be smarter. That's essentially what I hoped for from my TMS experience, except that I substituted emotional intelligence for reading proficiency.

The problem with that notion is that there will always be a top and a bottom in any range of function. We may turn a few people into super-readers, but will that benefit society? In another example, using brain stimulation to relieve the pain of autistic social isolation sounds great, but giving bright college students "emotional ESP" is not so obviously good. It certainly won't help the masses who don't get the therapy. When relieving disability takes away suffering we accept it, though my experience suggests that even that may come at a cost. Enhancing performance is not so obviously beneficial, yet there will be strong pressure to do it. A hundred years ago we imagined the prospect of improved humanity through eugenics, breeding the supermen of tomorrow. Soon, brain imaging and stimulation may offer us the ability to make ourselves into those supermen. But what will be the price?

Then there is what I think of as the security aspect of brain exploration. Medical imaging began as a way to see inside the body, but it's fast developing into a tool that can see patterns that represent our thoughts and feelings. The trouble is, reading those patterns does not tell you which ones a person will act upon. We often feel conflicting things, especially when something or someone is important to us. We may love and hate or be drawn and repelled at the very same moment.

Dr. Just's technology has obvious implications for the interrogation business. But is there really a circumstance in which such a tool can be used properly and ethically? That's an important question to discuss, because it may be inevitable that tools that evolve from work like his will be used to "understand" criminals and terrorists. The mind probe of science fiction could be a reality in 2026.

One thing that can help us resolve these looming ethical dilemmas is a closer connection between doctors, researchers, and the people affected by their work. Dr. Minshew showed a great example of that each time I visited her clinic. Whenever I went I saw bright researchers and fascinating projects. But the best part came after hours, with her group dinners. The first time we arrived at the restaurant I was surprised to find a diverse and motley crew of artists, musicians, activists, meditators, doctors, parents, researchers, and even a few financiers who support her research. The eclectic group was not what you'd expect from a staid research organization.

"I am focused on translating science into interventions," she told me, "testing the efficacy of new treatments and then seeing them disseminated. It's time for science to pay off on its promises." I couldn't agree more, but that's only part of the story. Nancy reminded me that you don't always need modern medicine to tap hidden reserves in the mind. Creative people are often different, and bringing them together can accomplish remarkable things.

Dr. Minshew was clearly at the center of a dynamic and thriving local autism community. Some of the autistic people she knew were severely disabled—even crippled by autism or its associated complications, at least in terms of their capacity to negotiate daily life. Yet everyone in attendance—me included—felt its mix of eccentricity and exceptionality and were empowered and excited by that.

I realized they were all working to make everyone's lives better—with scientists harnessing technology for tomorrow as clinicians, doctors, and natural practitioners worked to help today, informed by emerging knowledge. The artists and musicians played a valuable role, encouraging creativity, expression, and sharing in the community.

"They help keep me grounded," she said, and I can't help but think her own ethical choices will be informed by the wisdom of the group—most of whom have a personal stake in her work. I hope that other researchers can tap into similar resources in their own communities. There's no better place to discuss the ethics of disability, difference, and treatment choices than with the affected population. It's when they are excluded that we have problems.

With all of that in mind, I eagerly await the coming revolution in the treatment of brain disorders as these techniques and technologies converge. Psychiatrists already criticize our excessive reliance on drugs. Alvaro and Lindsay have shown that brain stimulation offers an alternative in some cases, and its applicability will likely expand with increased experience. In the future, neurologists like Alvaro and imaging experts like Marcel will work hand in hand with cognitive therapists, combining brain imaging and stimulation with new and existing therapies. They will also work with radiologists to identify and target locations that are implicated in dysfunction. As Nancy pointed out, when this is backed up with genetic testing and other tools we are headed for a future of truly individualized treatment for conditions that, until now, have been thought to be unchangeable.

I hope that this coming convergence will cause neurology, psychology, and psychiatry to begin merging into a new and more holistic discipline, directed at better brain function, and that we'll soon talk of brain health and "making brains stronger" just as we talk about building muscles and stamina for a marathon. Alvaro's Brain Fit Club at Beth Israel is one of the first places to begin doing that.

One idea that I've come back to throughout this book is the notion that my brain (or anyone else's) might differ from the brain of a typical person. But the truth is, there is really no such thing as a "typical" brain, because every human is atypical in some or many ways. The "neurotypical person" is a construct, established by scientists who need parameters by which to measure the disparate statistics of different individuals.

Last year, Dr. Just tried a new kind of experiment with interesting implications for what "neurotypical" and "different" mean. Marcel put thirty-four young adults in his fMRI scanner and asked them to imagine the following verbs: compliment, insult, adore, hate, hug, kick, encourage, and humiliate. Then he asked them to consider the verbs from the perspective of applying them to another person, as well as to themselves.

In previous experiments he had already identified characteristic brain activity patterns that were associated with subjects imagining each of these actions, but this time he introduced a new twist. Half of these new subjects were autistic; the other half were not. His findings were remarkable, to say the least. Dr. Just was able to separate the autistic from the non-autistic subjects just by comparing their responses to those verbs, and the difference that set them apart was striking. He observed that the autistic people had a different pattern of activation of the brain areas associated with emotional response. Yet they showed the same activation as the non-autistic subjects to the logical meaning. Not only were the autistic responses different, they were consistent. In other words, the pattern of emotional response was similar in all the non-autistics and similar (but different from the first group) in all the autistics.

Dr. Just cautioned me that this study only included thirty-four subjects, and a much larger study would be needed to draw population-wide conclusions. But with that caveat, his research may point the way to a noninvasive tool that can identify autistic brains through a brief session in an fMRI scanner. That alone would be a major breakthrough.

But there's more. . . .

The finding that his subjects had similar brain response patterns to those stimulations suggests an answer to that age-old question, "Is my perception of red the same as yours?" At the same time, the fact that the autistic subjects had a collectively different response in some areas shows that neurological difference (in this case, autism) may alter our perceptions in a very elemental way. That should be a potent piece of knowledge for any therapist.

Alvaro and Lindsay made a similar discovery in their own lab. They found that autistic people were more affected by a brief burst of TMS, and for a longer time, than non-autistic subjects. They interpreted that finding as evidence that autistic people have more plastic brains, a factor that could explain both gift and disability.

When your brain is too changeable in the parts that acquire basic skills, you won't be able to learn enough to function, because your brain will be unlearning whatever you are taught five minutes after you learn it. Yet a touch of extra plasticity could be an element of genius, if it helped your brain rewire itself for new abilities faster than an ordinary person's brain.

It will take us years to fully explore insights and issues like these. For every answer, we are presented with new questions. It's an exciting time!

Doctors will face more and more ethical dilemmas as they try to decide what should be treated and what might be left alone. The decision to treat brain damage after a stroke, for example, is fairly straightforward. But in other cases, "treatment" may become more about making patients aware of their differences and learning to see the aspects that are gifts. Who are we to say what's disabled and what's different, unique, and special in another person? As Howard Gardner first wrote decades ago in *Multiple Intelligences,* there are a variety of distinct intellectual capacities and orientations that contribute to our understanding of ourselves and our place in society.

Sometimes, a touch of disability is what makes us great. Consider a few historical figures who were touched by autism, serious eccentricity, or some other disability in addition to their well-recognized gifts:

Leonardo da Vinci
Michelangelo
Ludwig van Beethoven
Isaac Newton
Wolfgang Amadeus Mozart
Albert Einstein

Today, we might add to the list folks like Bill Gates and Dan Aykroyd.* Where would we be if we'd made those people "normal" in childhood? As Temple Grandin says, "We'd be living in caves and using our social skills to tell each other jokes by firelight." My vote is for a renaissance in which

---

* Cambridge autism researcher Simon Baron-Cohen and Trinity College professor Michael Fitzgerald have both written about historical figures who were likely autistic. In modern times, Bill Gates has been widely described as exhibiting many traits of autism, but he has not to my knowledge made a statement. Dan Aykroyd disclosed his Asperger's in an interview with the *Daily Mail* in December 2013.

psychology, psychiatry, neurology, and radiology (with the latest brain imaging techniques) come together to achieve unprecedented success in treating formerly untreatable neurological differences and disorders. The brain scientists I've been privileged to meet are truly pointing us toward a better tomorrow.

Afterword

Marcel Adam Just, PhD

THE STORIES THAT John Elder Robison tells in these pages come together in a fascinating journey, one that reveals facets of daily life that are often not evident to people who are not on the spectrum, told from the unique perspective of an autistic person. Perhaps the most dramatic part of this story is John's experience following his participation in a TMS study at Harvard and Beth Israel Hospital. That brief journey brings chills to one's spine as John describes in eloquent detail how his own thinking and perceptions temporarily changed after the experimental treatment. This is probably the most articulate description ever written of a phenomenological experience triggered by TMS. I am not sure what to make of it. I somewhat envy him the experience, and I would like to think that I would be brave enough to embark on a similar trip.

The part of John's journey that is closest to my heart describes his learning about contemporary cognitive neuroscience. Of course, that field is changing as I type this sentence, and it will surely have changed by the time anyone reads it. But it seems to me that understanding how we get mind from brain is one of the most interesting questions faced by humankind. How is it that those three pounds of tissues containing eighty-six billion neurons can invent a flying machine, write a poem, and raise a child? Those are remarkable achievements for such a small organ, one that we almost never see or even sense (given that it doesn't have pain receptors, for example). Consider the miracle that occurs in each pregnant mother as her unborn child's brain is gradually constructed according to the world's most complex blueprint. It is of course amazing that newborns come with beating hearts and working lungs, that they can eat and poop. But to me what's most amazing is that they also come with an organ of thought that can sense and respond to the world, and which can eventually invent drones, write essays, and develop web start-ups.

Cognitive neuroscience is the discipline that tries to explain how a brain gives rise to a mind. The brain-mind combo can get us through day-to-day life, and it can think deep thoughts. It governs how we perceive and interact with other people. It enables us to provide food and shelter for ourselves and those we care for. It is only a bit of an overstatement to say that human beings are brains in a body. Our personalities, skills, feelings, and knowledge all reside in our brains. That is why cognitive neuroscience is such a fundamental human science. It attempts to explain who we are. Brain imaging, particularly fMRI, has arguably taught us more about human brain function in the last twenty-five years than we ever previously knew.

The gift that imaging of brain *function* (the "f" in fMRI) has bestowed on us is a picture of brains at work. At first, in the 1990s, those fMRI pictures showed us the hot spots in our brain while a particular type of thinking task was being performed. Those early pictures provided several fascinating insights. They showed that every task evoked activation (a hot spot) not just in one area but in an entire set of brain areas, showing that thinking was a collaboration among ten to twenty brain centers, like a team sport among specialists whose work is intricately coordinated. The early fMRI pictures also showed that in people with neurological or psychiatric disorders, the activation pattern was often disordered. In people with high-functioning autism, the alteration was manifested as poorer synchronization between frontal brain areas and other brain areas. Because many types of thinking (particularly social processing) involve coordinated activity between frontal and other areas, this lower level of synchronization provides an explanation of the social thought alteration in autism.

A second brain imaging revolution occurred at the beginning of the twenty-first century, when new computational techniques, particularly the branch of computer science called machine learning, began to be applied to brain imaging data. I was fortunate to be part of this adventure in a collaboration with Tom Mitchell, a pioneer in machine learning, as well as with many extremely talented postdoctoral fellows, graduate students, and research staff. We set off to relate the patterns of brain activity to specific thoughts, and not just treat them as hot spots. This is the ongoing work that John refers to as mind reading. As John explains, it is now possible to tell what concept a person is thinking about from its brain activation signature. Our work started by identifying the signatures of concrete concepts, like the thought of an apple or a hammer or an igloo. We progressed to identifying experiences of emotions, so that we could tell whether someone was feeling happiness or disgust, for example. We also found, as John says, that thoughts of social interactions like hugging were altered in people with high-functioning autism. This discovery suggests that many if not all thought disorders may be amenable to diagnosis with our mind-reading techniques. And the way in which a thought is altered may indicate where to target a therapy.

It would be very satisfying if fMRI mind reading progressed from being the world's greatest parlor trick to being a tool that reveals the inner structure and workings of human thought. The most promising aspect is that not only can a concept's fMRI signature be detected, but it can also be decomposed into its main components, revealing the building blocks of the concept. The experimental fMRI findings illuminate which set of brain regions underpins each building block. As we discover the building blocks of more and more types of concepts, we hope to understand the nature of all types of thoughts, from the thought of an apple to the thought of "The old man threw the stone into the lake" (which is a thought of the complexity that we can currently read).

Aside from the satisfaction of revealing the brain's Lego pieces for building thoughts, this type of research has the potential to tangibly change our world in two main ways. First, as mentioned in the book and above, this approach has the potential to change our understanding and treatment of many types of thought disorders. Rather than simply knowing that some aspect of thought is disordered—say, for example, that a person has an exaggerated sense of being persecuted—it may become possible to say precisely which building block of thought is absent or misshapen and how the functioning of the underlying brain areas is altered. For example, someday, instead of saying "This child has dyslexia," we may say "the anatomical connections between two regions in the frontal cortex of this child are undermyelinated." With such a diagnosis, it could be possible to apply a therapy that improves the connectivity and improves the reading to near-normal levels. That "someday" has already occurred, and was described in the journal *Neuron* in 2009 in an article by Tim Keller and myself (available at ccbi.cmu.edu/publications.html). More generally, many thought disorders may eventually be traceable to their brain basis in a way that suggests a therapeutic intervention.

The second possible benefit of understanding the brain's building blocks is that it could enhance instruction processes in many types of education. If we know the desired end state of the thoughts of a learner, we might be able to design the instruction to optimally provide and assemble the building blocks that compose the thought. The desired end state can be determined (using "mind reading") from the brains of people who already have an excellent grasp of the end-state concepts. Such research is in its infancy, but it could revolutionize how we educate our young, train our workforce, and transform every child into a literate, thoughtful, productive citizen.

This vision sounds grandiose, but so would have been a description of the current state of knowledge about brain function had someone foreshadowed it twenty-five years ago. John Elder Robison has given us here his glimpse into the future. No one can accurately predict the future of brain science, but we can be certain that it will be here soon and that it will change the world.

MARCEL ADAM JUST
D. O. Hebb Professor of Psychology and
Co-director of the Center for Cognitive Brain Imaging,
Carnegie Mellon University

Findings and Further Reading

MANY READERS ASK ME where TMS therapy stands today, and whether they can sign up for treatment. As this book goes to press, in early 2016, TMS is available as an FDA-accepted therapy for depression at hundreds of hospitals and clinics across the United States. It's also available in Canada, Europe, Australia, and parts of Asia.

TMS is not yet an FDA-approved therapy for autism or ADHD, but I suspect that will come in the next decade. Right now, the scientists described in this book are forging ahead with new studies that build on what you've read in the previous pages. For those of you who want to follow in my footsteps, they are always looking for volunteers. I've focused on a few talented researchers in this book, but they are not the only ones who are active in the field. In America, Manny Casanova continues to study autism and TMS, and as of this writing, he has actually conducted more varied trials than anyone else in America. Peter Enticott is breaking new ground in his TMS center at Australia's Deakin University, and Simon Baron Cohen and others study TMS and autism at Cambridge in the UK. My website (johnrobison.com) contains a more up to date list of TMS centers and studies.

Researchers are also exploring the use of very low-power electrodes to deliver energy to the brain. That technology is called TDCS, for transcranial direct current stimulation. Some studies have shown it to have an effect similar to TMS, and it's being studied in Lindsay's lab right now. Other scientists have found that they can deliver energy to the cortex with certain frequencies of laser light that penetrate the cranium and energize the underlying brain tissue. The thing those techniques and TMS all have in common is that they are pure-energy therapies, and they all target small regions of the brain in a safe, noninvasive manner. Collectively, they are one of the biggest—and least recognized—advances to ever happen in the field of neuroscience.

Manny Casanova is studying the convergence of TMS and neurofeedback. By stimulating regions of the brain and using neurofeedback to optimize brain-wave patterns, he hopes to make bigger advances than TMS alone would give. One day we might see a triad of therapies—brain stimulation, cognitive therapy, and neurofeedback—all combined into one powerful package. When guided by next-generation brain imaging, the results may be beyond our wildest dreams.

Manny has agreed to write some thoughts on that, and a further elaboration of how he thinks TMS works in a piece on my website. I encourage you to look there for more information and commentary.

Most of this research is being done in labs at university medical centers, but there are also independent doctors experimenting in private clinics. Even though their treatments are not presently approved by the Food and Drug Administration and they are not yet validated by peer-reviewed studies, they may still be life-changing, as my story and the stories of others attest. It's my opinion that TMS is inherently safer than medication. Even though medications are tested extensively before release to the public, we continue to find unexpected chemical interactions, sometimes decades later. As a pure-energy treatment, TMS doesn't add chemicals to the body. Consequently, its potential for those kinds of side effects is near zero. TMS is finding increasingly broad acceptance in the treatment of depression, and its side effects there are usually described as markedly less than those from psychiatric medications. Still, as a prospective patient you have to make your own judgment about the skill and capability of the doctor offering TMS therapy, just as I did. Review boards oversee doctors who work at hospitals like Beth Israel; independent practitioners generally are not monitored very closely, so it's important to check reputations.

When I talk to doctors or psychologists about TMS, I'm always asked about findings. It's great to hear your story, they tell me, but what did the peer-reviewed journal accounts say?

In July 2011, Shirley, Lindsay, Alvaro, and others involved in the research published the results of the first TMS study in the *European Journal of Neuroscience,* under the heavy title "Brain Stimulation over Broca's Area Differentially Modulates Naming Skills in Neurotypical Adults and Individuals with Asperger's Syndrome." The three-year lag between our participation in the TMS lab and publication of the results is typical in medical research.

Reading their paper feels funny, because it's a very dry account of what was a very emotional and transformative time for me, and I'm sure for the other subjects. Our experiences in the hours and weeks after TMS are not mentioned at all; the published findings are limited to the results of the before and after testing in the lab, and an analysis of what they may mean.

While I can't argue with the scientific validity of that approach, it does not come close to describing everything that went down. Here's how the scientists summed it up:

> Object naming was assessed before and after low-frequency rTMS of the left pars opercularis, left pars triangularis, right pars opercularis and right pars triangularis, and sham stimulation, as guided stereotaxically by each individual's brain magnetic resonance imaging. In ASP participants, naming improved after rTMS of the left pars triangularis as compared with sham stimulation, whereas rTMS of the adjacent left opercularis lengthened naming latency. In healthy subjects, stimulation of parts of Broca's area did not lead to significant changes in naming skills, consistent with published data.

Overall, these findings support our hypothesis of abnormal language neural network dynamics in individuals with ASP.

But let me attempt to explain what all of that means, and what happened, in plain English. For the first TMS study, we were shown objects on a computer screen, and we had to name them as quickly as we could. It went something like this:

Bird
Stethoscope
Car
Protractor

Some of the objects were simple to name; others were not so easy. At the beginning of the study we were asked to name seventy-five objects, which were then broken down into five lists of fifteen. After each TMS session we were shown one of the lists and our responses were recorded and evaluated.

There were ten of us on the autism spectrum (all diagnosed with Asperger's syndrome and most having above-average IQ) paired up with ten similar folks who weren't autistic. Each of us did the tests alone, and for the most part, we did not meet or talk with one another. Interestingly, our performance on the naming test was not materially different at baseline. Differences appeared only after TMS, with one stimulation making us do better and another making us do worse. When stimulated in the left pars triangularis, those of us in the Asperger's group outperformed the controls, who did not change. When our left pars opercularis was stimulated our performance got worse, and again the controls did not change. The stimulations on the right side of the brain didn't affect the naming results at all, but they had a huge impact on several of us with respect to emotional insight.

It's interesting that the most profound effect of the study—that TMS stimulation of the right forebrain could unlock emotion in some of us—was not mentioned at all, because that result was not an original study objective. Neuroscience papers follow a fairly rigid form, where the experimenters describe a hypothesis to be tested, the test method, and the results. There's little place for the unexpected surprise.

It's funny . . . you could read that paper and have no idea it was based on the same TMS sessions I describe in this book. Or you might read that paper after reading this book and conclude that the original objectives of the study were rendered irrelevant. In that light, you might say this book describes the most important findings of those studies, at least from my perspective.

Lindsay published another TMS study entitled "Abnormal Modulation of Corticospinal Excitability in Adults with Asperger's Syndrome" in the *European Journal of Neuroscience* in September 2012.

In that study, which ran concurrently with the other TMS studies I took part in, Lindsay stimulated the motor cortex and made our fingers jump as I described earlier in this book. Next she applied a burst of depressive TMS, which suppressed our fingers' tendency to twitch when stimulated by single pulses as she'd done at the outset. Then Lindsay measured how long it took for the effect of that depressive burst to wear off.

Those results were fascinating because they showed a striking difference between the autistic and non-autistic participants. Most members of the latter group recovered finger neuron response within fifteen minutes. In contrast, the autistic people took two to three times as long to recover. As Alvaro later explained to me, that suggests that the behavioral and emotional effect of TMS may last *long* in autistic people—in fact long enough for TMS to change us in a lasting manner by reinforcing plastic changes, just as some of us experienced.

It is also meaningful from a safety point of view. If people "just try" TMS in unsupervised or uninformed experimentation, it might permanently modify behavior in ways that could be undesirable.

By looking at whether a person's response had recovered fifty minutes after the TMS burst, Lindsay was able to distinguish autistic and non-autistic individuals with an accuracy of 93 percent. That's better than most diagnostic tests in use today, and it suggests that excess plasticity might be a distinguishing feature of the autistic brain.

The issues surrounding plasticity in autism are so important that the National Institutes of Health have funded Alvaro's lab to study them through 2019.

If you'd like to know more about Alvaro's theory, you can read his chapter "The Metamodal Organization of the Brain," in *Progress in Brain Research,* vol. 134, edited by C. Casanova and M. Ptito (New York: Elsevier, 2001), available on the TMS lab website: http://tmslab.org/includes /alvaro_3.pdf.

One study Nick took part in was published in August 2014 in *Frontiers in Human Neuroscience,* with the daunting title of "Modulation of Corticospinal Excitability by Transcranial Magnetic Stimulation in Children and Adolescents with Autism Spectrum Disorder."

The underconnectivity theory that Dr. Just and Dr. Minshew told me about was published as "Autism as a Neural Systems Disorder: A Theory of Frontal-Posterior Underconnectivity" in *Neuroscience & Biobehavioral Reviews,* February 6, 2012.

Dr. Just's ideas on TMS were published as "Neurocognitive Brain Response to Transient Impairment of Wernicke's Area" in the journal *Cerebral Cortex,* January 14, 2013.

Dr. Just's study that identified autistic people on the basis of brain response was published as "Identifying Autism from Neural Representations of Social Interactions: Neurocognitive Markers of Autism" in *PLoS One,* December 2014.

There is an extensive index of Dr. Just's other writings and papers on his website at psy.cmu.edu/people/just.html.

Dr. Minshew's university profile page can be found at psychiatry.pitt .edu/node/7900.

What about the autism science work I've done? All the output of the Interagency Autism Coordinating Committee is available for free download from the government website iacc.hhs.gov. We produced annual reports each year in which we discussed advances and achievements and also identified questions for the future. Our meetings are also archived online. If you look at the autism sections of the National Institutes of Health and Centers for Disease Control websites you'll find more material, including my talks before both groups during April, which is Autism Acceptance Month.

One of the activities that I'm involved with now is the development of ICF Core Sets for Autism for the World Health Organization. Our group has published several papers, including:

"Classification of Functioning and Impairment: The Development of ICF Core Sets for Autism Spectrum Disorder," *Autism Research* 7, no. 1 (2014): 167–72.

"Ability and Disability in Autism Spectrum Disorder: A Systematic Literature Review Employing the International Classification of Functioning, Disability and Health—Children and Youth Version," *Autism Research,* in press; published online March 2015.

A Google search will turn up videos and texts of my talks at IMFAR, the International Meeting for Autism Research. Every year, I speak at the conference to provide an autistic person's perspective. I can't imagine doing something like that before TMS.

The 2015 IMFAR conference took place in Salt Lake City, and I was one of the keynote panelists. In an impromptu talk, I spoke about autism, disability, and how much science and scientists had meant to my life. Difference may look like a gift, but it can be very painful to live, and science as I describe it in this book has helped me more than I can ever say. I believe my story of the personal impact of autism research came as a surprise to some of the audience, who gave me a standing ovation—the only one in IMFAR history, I was told. And, similar to the night I spoke of earlier in this book, when I addressed the doctors of my local medical society, there was not a dry eye in the house. That same thing happened seven years later at IMFAR. Connecting to the souls of others through their emotions is a powerful thing.

As this book goes to press, the International Society for Autism Research (INSAR), which hosts the IMFAR conference, is preparing to put that talk online for public viewing and it will be linked from my website.

Acknowledgments

IN THIS BOOK, I've done my best to translate some very complex science into something the average person can understand. I could never have done that without the help of the people listed in this section. When you read this list of credits, consider how all the people are tied together with multiple threads of connectivity, just like the strands in our brains. . . .

The first person I would like to thank is former Elms College president James H. Mullen, Jr. Dr. Mullen is now president of Allegheny College and chairman of the American Council on Education. He got this story started by inviting me to the Elms, where Dr. Kathryn James welcomed me into their autism program. If not for them, I'd never have met Lindsay. . . .

I will always be indebted to Dr. Alvaro Pascual-Leone, his postdocs Lindsay Oberman, Ilaria Minio-Paluello, Shirley Fecteau, and all the other staff of the Berenson-Allen Center for Noninvasive Brain Stimulation at Boston's Beth Israel Deaconess Medical Center. They are all bright, compassionate, and driven researchers. Alvaro is still very active in TMS research, and his clinic has developed a significant therapeutic practice in Boston. He remains one of the world's leading TMS researchers.

In addition to heading the TMS lab, Alvaro is a professor in Neurology and the associate dean for clinical and translational research at Harvard Medical School.

Lindsay's work earned her a lab of her own. Today she is director of the Neuroplasticity and Autism Spectrum Disorder Program at Bradley Hospital, the scientific director of the clinical TMS program at Rhode Island Hospital, and an assistant professor in the Department of Psychiatry at the Warren Alpert Medical School of Brown University in Providence, Rhode Island. She works with TMS and autism and collaborates with Alvaro and others at Beth Israel, Boston Children's, and Harvard.

Shirley moved back to Canada after her postdoc at Harvard. She currently holds a Canada Research Chair in Human Cognition, Decision-making, and Brain Plasticity at Laval University in Quebec. Much of her work today is focused on the use of TMS and other noninvasive brain stimulation techniques to help people with addictions.

Ilaria returned to Rome after her postdoc, where she joined the faculty of Sapienza University. She continues to be interested in TMS and autism, and I was pleased to visit her in Rome with my wife, Maripat, in the fall of 2012.

After getting involved with those Harvard scientists, I relied on my radiologist friend David Rifken, MD, to explain MRI and other brain imaging and the strange dots on my brain scans, and to answer a million other bizarre questions. Dave is a senior radiologist at Cooley Dickinson Hospital of Northampton, a Massachusetts General Hospital affiliate. Thanks again for being my friend and reading my brain scans (and all the others over the years!).

Thanks to Kimberley Hollingsworth Taylor, Michael, Nick, and family for accompanying me on the TMS journey, and for contributing memories to this account.

I'd also like to thank Dr. Nancy Minshew and Dr. Marcel Just at the University of Pittsburgh Autism Center and Carnegie Mellon University. The work they are doing is truly mind expanding and I cannot thank you enough for inviting my wife and me to be part of it.

Dr. Manny Casanova and his wife, Dr. Emily Casanova, are important people in the world of TMS and (at the time of my TMS experiences) at the University of Louisville. Both played a big role in the background by helping me understand TMS and the workings of the mind, and I greatly appreciate their contributions. Manny now holds the SmartState Endowed Chair in Translational Neurotherapeutics. He's a professor of biomedical sciences at the University of South Carolina School of Medicine Greenville.

He is also the closest thing I know to a Renaissance man of neuroscience. While I describe researchers like Lindsay as users of TMS technology, Manny is a guy who does not hesitate to roll up his sleeves and go into the guts of the machine to get what he wants from it. As I've gotten to know him these past six years, I've often wished we lived closer together. He agreed, saying, "We would have enjoyed spending some time together building our own TMS machine with a bank of super capacitors, trying not to shock ourselves, and examining the resultant brain-wave changes as if they were on an oscilloscope."

Peter Enticott has also weighed in on this story, and I'd like to recognize his contribution here. Peter helped out in the background for the American edition of this book, and wrote a foreword for the Down Under edition. He's a professor of cognitive neuroscience at Deakin University in Melbourne, Australia, and another top TMS scientist.

Next I'd like to thank Dr. Geri Dawson for asking me to get involved in autism science and then serving as a guide and mentor as I found my way. When I began my NIH service I was supported by everyone I met—from National Institute of Mental Health director Dr. Thomas Insel, to National Institute of Child Health and Human Development director Dr. Alan Guttmacher, Office of Autism Research Coordination head Dr. Susan Daniels, and all the members of the Interagency Autism Coordinating Committee with whom I've had the honor of serving since 2012.

At the Centers for Disease Control and Prevention I was encouraged by Dr. Marshalyn Yeargin-Allsopp, Dr. Cathy Rice, and many others. Whenever I have a question you are always there, and I hope I've been able to reciprocate by my work on your committees.

Five years ago, Geri suggested I offer an autistic person's perspective to the scientists of INSAR, the International Society for Autism Research.

Today I am a voting member of INSAR and a member of its governing committees. I speak up for the role of autistic people in guiding research for and about us. I'd like to express my thanks to all of the INSAR members who have spent time talking to me and encouraging me on this journey.

Lisa Greenman deserves special thanks for connecting me to the Ivymount School in 2008. Lisa is a federal public defender with a focus on defending people with developmental disabilities. It was she who counseled me to speak up for autistic people who are unfairly charged with crimes by ignorant or misguided prosecutors.

Monica Adler Werner at the Ivymount School showed me what an autism school can be, as did Dr. Marty Webb of Monarch School in Houston, Texas. Both of you have inspired me more than you know, with the evidence being the high school program in our car complex today.

Monica turned out to be involved with a team of researchers at the Children's National Medical Center's Center for Autism Spectrum Disorders who wanted to find practical ways to help kids on the spectrum. "Medical research is fine," she told me, "but we still have to teach kids how to get by in the world right now." It turned out they were developing a curriculum called Unstuck and testing it in classrooms at her school. What I saw was so impressive that I wrote a foreword for the *Unstuck and On Target!* book, and I recommend it wherever I go. Next year, we hope to begin testing the Unstuck techniques at William & Mary.

I'd like to thank everyone at the College of William & Mary, particularly history professor Karin Wulf, who first brought me there, provost Michael Halleran, who believed in neurodiversity at a time when the word was unknown in higher education. Thanks to Joel Schwartz, director of the Charles Center and dean of honors and interdisciplinary studies, for welcoming me into his college. I'd also like to recognize my other colleagues in the neurodiversity program—Professors Josh Burk, Cheryl Dickter, Warenneta Mann, and Janice Zeman. I'm proud to be associated with all of you. Thanks also to the Aronow Foundation for their support of W&M's neurodiversity effort.

Thanks to my old friend Eugene Cassidy, president of the Eastern

States Exposition, and all his crew—especially John "JJ" Juliano and Gerard Kiernan—for providing such a wonderful venue for photography and a uniquely colorful cast of characters.

Thanks to Maribeth, Martha, and the staff of Robison Service. All of you keep the place running while I am engaged in these projects.

I owe a debt of gratitude to Steve Ross for believing in me over the years. It was Steve—as the head of Crown in 2007—who initially brought me into Random House when he saw the manuscript that became my first book. Now, with Steve as my literary agent, I've remained in the Random House family with editor Jessica Sindler and publishers Cindy Spiegel and Julie Grau. For those of you who wonder what editors and publishers do, they took the jumble of words from my neurodiverse mind and arranged them to flow smoothly, as you read them today. While every word in this book is my own, my editor and publisher were instrumental in putting them into the fine form you read today. They eliminated repetitions, corrected strange styles, and moved passages around so the story marched ahead as smoothly as the beat of the TMS machine. You might say they arranged my words so the print on these pages matched the vision of the book in my mind—a remarkable feat!

I would also like to recognize the efforts of a few others on the Random House team: publicity director Sally Marvin, deputy publisher Tom Perry, art director Greg Mollica, production editor Beth Pearson, copy editor Amy Morris Ryan, and Amelia Zalcman in the Random House legal department.

Finally I owe a great debt to my lovely wife, Maripat. She has shown me a family life I'd never known, and she's beside me in all that I do. She feeds, nurtures, and cares for me despite my difficult ways, and she tolerates the things I do with mostly good humor. As I said in the dedication, she's knitted our blended family together with love.

Thanks also to the rest of our family—Julian, Joe, Jack, and Lindsay—for putting up with me and believing in this project. There have been many nights that I sat upstairs writing till midnight when I could have been entertaining our pets, playing board games, tending to household responsibilities, or doing other more family-like activities.

This is without a doubt the hardest book I have written. The many different characters and the multiple threads of the story made the process quite complex. When characters got up and moved on their own I found it so unnerving that I would stop writing for days. Glue eventually held them down. Anytime tools were needed I used pliers from Knipex and screwdrivers from Snap-on. Bosch solar cells gathered the energy to write by, and pulp lumber for the page proofs came from our forest. Getting the finished work from the wilds of Massachusetts to Random House in New York was a difficult challenge, but my five-ton army cargo truck was up to the task. As I told the startled doorman at 1745 Broadway, ten-wheel drive will take you anywhere. My uncle Mercer showed me that forty years ago, and I never forgot. Remember that well in your own travels.

JOHN ELDER ROBISON is a world-recognized authority on life with autism, and the *New York Times* bestselling author of *Look Me in the Eye, Be Different,* and *Raising Cubby*. Robison is the Neurodiversity Scholar in Residence at the College of William & Mary, and he serves on the Interagency Autism Coordinating Committee, which produces the U.S. government's strategic plan for autism spectrum disorder research. A machine aficionado and avid photographer, Robison lives with his family in Amherst, Massachusetts.

johnrobison.com
Facebook.com/JohnElderRobison
@johnrobison

Printed in the United States
by Baker & Taylor Publisher Services